Combined Workshop on Spatial Language Understanding (SpLU) and Grounded Communication for Robotics (RoboNLP) 2019

Minneapolis, Minnesota, USA

6 June 2019

ISBN: 978-1-5108-8770-1

NAACL HLT 2019

The Combined Workshop on Spatial Language Understanding (SpLU) and Grounded Communication for Robotics (RoboNLP)

Proceedings of the Workshop

June 6, 2019
Minneapolis, MN

Introduction

SpLU-RoboNLP 2019 is a combined workshop on spatial language understanding (SpLU) and grounded communication for robotics (RoboNLP) that focuses on spatial language, both linguistic and theoretical aspects and its application to various areas including and especially focusing on robotics. The combined workshop aims to bring together members of NLP, robotics, vision and related communities in order to initiate discussions across fields dealing with spatial language along with other modalities. The desired outcome is identification of both shared and unique challenges, problems and future directions across the fields and various application domains.

While language can encode highly complex, relational structures of objects, spatial relations between them, and patterns of motion through space, the community has only scratched the surface on how to encode and reason about spatial semantics. Despite this, spatial language is crucial to robotics, navigation, NLU, translation and more. Standardizing tasks is challenging as we lack formal domain independent meaning representations. Spatial semantics requires an interplay between language, perception and (often) interaction.

Following the exciting recent progress in visual language grounding, the embodied, task-oriented aspect of language grounding is an important and timely research direction. To realize the long-term goal of robots that we can converse with in our homes, offices, hospitals, and warehouses, it is essential that we develop new techniques for linking language to action in the real world in which spatial language understanding plays a great role. Can we give instructions to robotic agents to assist with navigation and manipulation tasks in remote settings? Can we talk to robots about the surrounding visual world, and help them interactively learn the language needed to finish a task? We hope to learn about (and begin to answer) these questions as we delve deeper into spatial language understanding and grounding language for robotics.

We accepted 8 archival submissions and 12 cross-submissions.

Organizers:

James F. Allen, University of Rochester, IHMC
Jacob Andreas, Semantic Machines/MIT
Jason Baldridge, Google
Mohit Bansal, UNC Chapel Hill
Archna Bhatia, IHMC
Yonatan Bisk, University of Washington
Asli Celikyilmaz, Microsoft Research
Bonnie J. Dorr, IHMC
Parisa Kordjamshidi, Tulane University / IHMC
Matthew Marge, Army Research Lab
Jesse Thomason, University of Washington

Program Committee:

Malihe Alikhani, Rutgers University
Yoav Artzi, Cornell University
Jacob Arkin, University of Rochester
John A. Bateman, Universität Bremen
Mehul Bhatt, Örebro University
Jonathan Berant, Tel-Aviv University
Raffaella Bernardi, University of Trento
Steven Bethard, University of Arizona
Johan Bos, University of Groningen
Volkan Cirik, CMU
Guillem Collell, KU Leuven
Joyce Chai, Michigan State University
Angel Chang, Stanford University
Simon Dobnik, CLASP and FLOV, University of Gothenburg Sweden
Ekaterina Egorova, University of Zurich
Zoe Falomir, Universität Bremen
Daniel Fried, UCSF
Lucian Galescu, IHMC
Felix Gervits, Tufts
Hannaneh Hajishirzi, University of Washington
Casey Kennington, Boise State University
Jayant Krishnamurthy, Semantic Machines
Stephanie Lukin, Army Research Laboratory
Chris Mavrogiannis, Cornell
Dipendra Misra, Cornell University
Marie-Francine Moens, KU Leuven
Ray Mooney, University of Texas
Mari Broman Olsen, Microsoft
Martijn van Otterlo, Tilburg University, The Netherlands
Aishwarya Padmakumar, UT Austin
Natalie Parde, University of Illinois Chicago
Ian Perera, IHMC

James Pustejovsky, Brandeis University
Preeti Ramaraj, University of Michigan
Siva Reddy, Stanford
Kirk Roberts, The University of Texas
Anna Rohrbach, UC Berkeley
Marcus Rohrbach, FAIR
Manolis Savva, Princeton University
Jivko Sinapov, Tufts
Kristin Stock, Massey University of New Zealand
Alane Suhr, Cornell
Clare Voss, ARL
Xin Wang, University of California Santa Barbara
Shiqi Zhang, SUNY Binghamton
Victor Zhong, University of Washington

Invited Speakers:

Dhruv Batra, GaTech/FAIR
Joyce Chai, Michigan State University
Cynthia Matuszek, UMBC
Raymond J. Mooney, UT Austin
Martha Palmer, CU Boulder
Matthias Scheutz, Tufts
Stefanie Tellex, Brown
Dilek Hakkani-Tur, Amazon

Table of Contents

Workshop Program

Thursday, June 06, 2019

08:30–08:40   *Opening Remarks*
Workshop Chairs

08:40–12:30   *Morning Session*

08:40–09:00   *Poster Spotlight (1 min madness)*

09:00–12:30   *Morning Session*

09:00–09:45   *Invited Talk*
Joyce Chai

09:45–10:30   *Invited Talk*
Matthias Scheutz

11:00–11:45   *Invited Talk*
Martha Palmer

11:45–12:30   *Invited Talk*
Stefanie Tellex

12:30–14:00   ***Session Poster: Poster Session and Lunch***

Corpus of Multimodal Interaction for Collaborative Planning
Miltiadis Marios Katsakioris, Helen Hastie, Ioannis Konstas and Atanas Laskov

¿Es un plátano? Exploring the Application of a Physically Grounded Language Acquisition System to Spanish
Caroline Kery, Francis Ferraro and Cynthia Matuszek

From Virtual to Real: A Framework for Verbal Interaction with Robots
Eugene Joseph

Learning from Implicit Information in Natural Language Instructions for Robotic Manipulations
Ozan Arkan Can, Pedro Zuidberg Dos Martires, Andreas Persson, Julian Gaal, Amy Loutfi, Luc De Raedt, Deniz Yuret and Alessandro Saffiotti

Multi-modal Discriminative Model for Vision-and-Language Navigation
Haoshuo Huang, Vihan Jain, Harsh Mehta, Jason Baldridge and Eugene Ie

Semantic Spatial Representation: a unique representation of an environment based on an ontology for robotic applications
Guillaume Sarthou, Aurélie Clodic and Rachid Alami

SpatialNet: A Declarative Resource for Spatial Relations
Morgan Ulinski, Bob Coyne and Julia Hirschberg

What a neural language model tells us about spatial relations
Mehdi Ghanimifard and Simon Dobnik

14:00–17:30   *Afternoon Session*

14:00–14:45   *Invited Talk*
Dhruv Batra

14:45–15:30   *Invited Talk*
Cynthia Matuszek

16:00–16:45   *Invited Talk*
Raymond Mooney

16:45–17:15   ***Best Paper Oral Presentations***

17:15–18:00   ***Continued Poster Session***

Corpus of Multimodal Interaction for Collaborative Planning

Miltiadis Marios Katsakioris[1], Atanas Laskov[2], Ioannis Konstas[1], Helen Hastie[1]
[1]School of Mathematical and Computer Sciences
Heriot-Watt University, Edinburgh, UK
[2]SeeByte Ltd, Edinburgh, UK
`mmk11, i.konstas, h.hastie@hw.ac.uk`
`atanas.laskov@seebyte.com`

Abstract

As autonomous systems become more commonplace, we need a way to easily and naturally communicate to them our goals and collaboratively come up with a plan on how to achieve these goals. To this end, we conducted a Wizard of Oz study to gather data and investigate the way operators would collaboratively make plans via a conversational 'planning assistant' for remote autonomous systems. We present here a corpus of 22 dialogs from expert operators, which can be used to train such a system. Data analysis shows that multimodality is key to successful interaction, measured both quantitatively and qualitatively via user feedback.

1 Introduction

Our goal is to create a collaborative multimodal planning system in the form of a conversational agent called VERSO using both visual and natural language interaction, which in our case will be images of the plan and messages. In this work, we focus on the domain of Autonomous Underwater Vehicles (AUVs). Experts in this domain typically create a plan for vehicles using a visual interface on dedicated hardware on-shore, days before the mission. This planning process is complicated and requires expert knowledge. We propose a 'planning assistant' that is able to encapsulate this expert knowledge and make suggestions and guide the user through the planning process using natural language multimodal interaction. This, we hope, will allow for more precise and efficient plans and reduce operator training time. In addition, it will allow for anywhere access to planning for in-situ replanning in fast-moving dynamic scenarios, such as first responder scenarios or offshore for oil and gas.

2 Previous work

Conversational agents are becoming more widespread, varying from social (Li et al., 2016) and goal-oriented (Wen et al., 2017) to multimodal dialog systems, such as the Visual Dialog Challenge (Das et al., 2017) where an AI agent must hold a dialog with a human in natural language about a given visual content. However, for systems with both visual and spatial requirements, such as situated robot planning (Misra et al., 2018), developing accurate goal-oriented dialog systems can be extremely challenging, especially in dynamic environments, such as underwater.

The ultimate goal of this work is to learn a dialog strategy that optimizes interaction for quality and speed of plan creation, thus linking interaction style with extrinsic task success metrics. Therefore, we conducted a Wizard of Oz (WoZ) study for data collection that can be used to derive reward functions for Reinforcement Learning, as in (Rieser, 2008).

Similar work is shown in (Kitaev et al., 2019), where the task involves two humans collaboratively drawing objects with one being the teller and the other the person who draws. The agents must be able to adapt and hold a dialog about novel scenes that will be dynamically constructed. However, in our scenario the agent must be capable of not only adapting but also identifying and editing specific attributes of the dynamic objects that are being created in the process.

Previous data collection on situated dialog, such as the Map Task Corpus (Anderson et al., 1991), tackle the importance of referencing objects while giving instructions on a drawn map with landmarks either for identification purposes or for displaying the perceived understanding of their shared environment. Our task is different in that it involves subjects collaboratively creating a plan

Proceedings of SpLU-RoboNLP, pages 1–6
Minneapolis, MN, June 6, 2019. ©2019 Association for Computational Linguistics

on a nautical chart rather than passively following instructions. In addition, our environment is dynamic. New objects are being created and the user with the agent, together, come up with the desired referring expressions (see Figure 3). A similar interactive method is described in (Schlangen, 2016), where they ground non-linguistic visual information through conversation.

In situated dialog, each user can perceive the environment in a different way, meaning that referring expressions need to be carefully selected and verified, especially if the shared environment is ambiguous (Fang et al., 2013). Our contributions include: 1) a generic dialog framework and the implemented software to conduct multiple wizard WoZ experiments for multimodal collaborative planning interaction; 2) available on request, a corpus of 22 dialogs on 2 missions with varying complexities and 3) a corpus analysis (Section 4) indicating that incorporating an extra modality in conjunction with spatial referencing in a chatting interface is crucial for successfully planning missions.

3 Method and Experiment Set-up

Our 'planning assistant' conversational agent will interface with planning software called SeeTrack provided by our industrial partner SeeByte Ltd. SeeTrack can run with real AUVs running SeeByte's Neptune autonomy software or in simulation and allows the planning of missions by defining a set of objectives with techniques described in (Lane et al., 2013; Miguelanez et al., 2011; Petillot et al., 2009). These can include, for example, searching for unexploded mines by surveying areas in a search pattern, while collecting sensor data and if, for example, a suspect mine is found then the system can investigate a certain point further (referred to as target reacquisition).

We used two wizards for our experiment, see Figure 1 for the set-up. We refer here to the wizards as 1) Chatting-Wizard (CW), who alone communicates with the subject getting information that is required to create the plan; and 2) the SeeTrack Wizard (SW), who sits next to the CW and implements the subject's requirements into a plan using SeeTrack and passes plan updates in the form of images to the CW to pass onto the subject. The subject was in a separate room to the wizards and interacted via a chat window for receiving text and images of the updated plan and sending text.

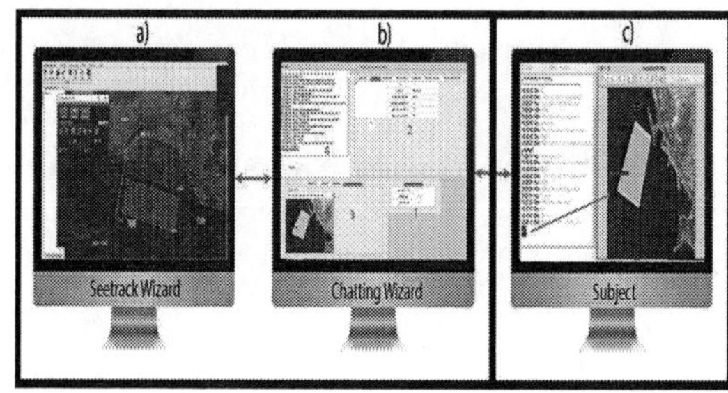

Figure 1: Experimental Set-Up, where a) SeeTrack Wizard, b) Chatting Wizard, and c) Subject console. Figure contains images from Seebyte's SeeTrack interface.

In order to establish the main actions and dialog act types for the system to perform, we recorded an expert planning a mission on SeeTrack whilst verbalizing the planning process and his reasoning. Similar human-provided rationalisation has been used to generate explanations of deep neural models for game play (Harrison et al., 2018). After analysing the expert video, we implemented a multimodal Wizard of Oz interface that is capable of sending messages, either in structured or free form as well as images of the plan. The Wizard Interface is made up of four windows (see Figure 2). The first has all the possible dialog acts (DA) the wizard can use together with predefined utterances for expedited responses. Once the DA is selected the predefined text appears in the chat window, from there the CW is able to modify as needed. The third window allows the CW to insert values (also referred to as 'slots') needed for the plan obtained through interaction from the user. Finally, the fourth window is for recording session details such as subject ID. The CW works collaboratively with the subject to develop a list of the necessary parameters that the SW needs to create the plan.

Each subject was given a short questionnaire to collect demographic information and instructions on how to approach the task of planning a mission using a conversational agent. A mission in our context is comprised of a nautical chart and a description of some objective that the subjects, together with the wizards, have to achieve. There are two main categories of missions A and B. The first (A) involves sending AUVs to survey areas of interest on the chart and is more time consuming. The second (B) category entails the reacquisition of a target, which overall can be achieved in less

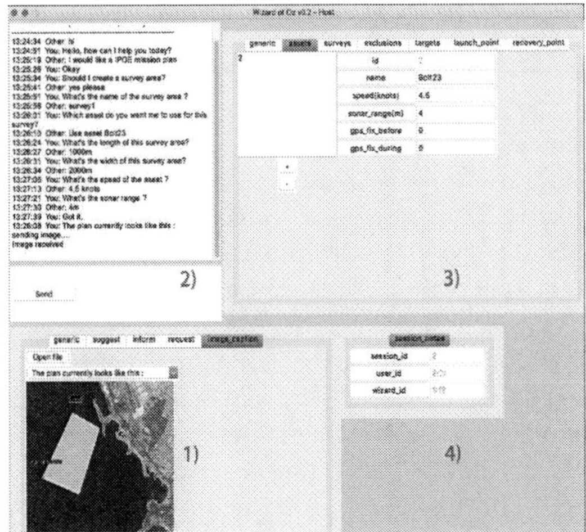

Figure 2: Wizard of Oz interface (Figure 1b) used by the Chatting Wizard, 1) dialog acts and structured prompts, 2) Chatting window, 3) State of the plan in the form of value-slots, 4) Session notes. Figure contains images from Seebyte's SeeTrack interface.

time. This is because surveying an area requires extra interactions between the operators and typically more spatial commands. Subjects were told that they had to plan both missions within a total time of one hour. Table 1 shows that the first mission took more time to plan in terms of actual planning time and number of user/system turns. This is normal because the first mission was of category A, which is by design harder, and the second mission of category B. The wording of the mission was very high level, so as not to prime the subjects. There are many elements that make up a plan (e.g. survey areas, exclusion zones) and therefore many variations are possible. Once both missions were completed, a post-questionnaire was administered to obtain their subjective opinion on the planning process. Subjects were told at the end that they were interacting with a wizard.

3.1 Subject Group

Planning missions for AUVs is a complex task, especially in the case of sophisticated software, such as SeeTrack. For this reason, we decided to focus our study mostly on expert users, who are familiar with SeeTrack and AUVs. We recruited human operators from industry (10 male, M_{age}=30), who all had some experience with SeeTrack.

4 Corpus Analysis

We collected 22 dialogs between the wizards and the subjects, which were analysed by a single annotator (an author). Figure 3 shows an example of a dialog interaction with corresponding dialog acts. We split our analysis into objective and subjective measures.

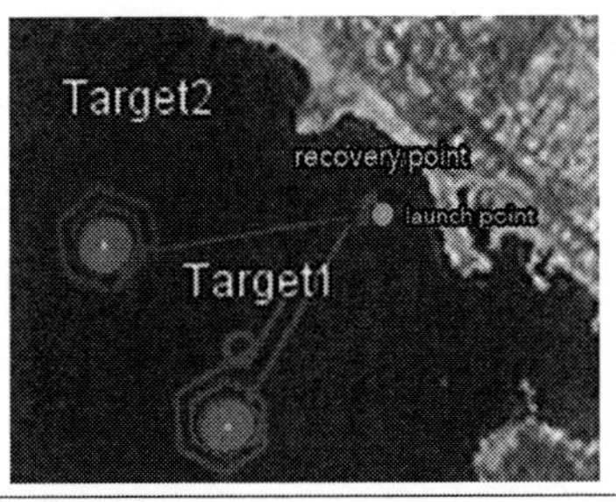

1. USER: 'move t2 200m west of t1'[inform]

2. SYSTEM:'Could you repeat that in different words? t1? t2?'[repeat]

3. USER:'move target2 200m west of target1'[inform]

Figure 3: Dialog excerpt and the corresponding image, displaying 2 targets, a launch and a recovery point [dialog act]. Figure contains images from Seebyte's SeeTrack interface.

4.1 Objective Measures

Dialog act types were adopted from the ISO (24617-2:2012) standard for dialog act annotation. Figure 4 gives the distribution of dialog acts, which were categorized into five groups:

1. **Generic** (conversational acts): wait, ack, affirm, yourwelcome, thankyou, bye, hello, repeat, praise, apology

2. **Inform** (for informing of values for slots): inform, negate, delete, create, correction, plan_complete, plan_mission

3. **Request** (for requesting information): request, enqmore

4. **Suggest** (for making suggestion): suggest

5. **Image** (for interacting with images): image_caption, show_picture

The most frequent user DA is the "inform" dialog act (54%), which informs the system about the plan slot values. This dialog act is also used for

Measures	Mission 1	Mission 2
# of turns	26.4(9.1)	13.1(4.4)
# of system utterances	51.4(21.0)	27.0(7.5)
# of user utterances	36.4(14.4)	19.7(8.1)
# of produced images	8.8(3.8)	5.0(1.3)
Time-on-Task (min)	26.3(0.005)	14.5(0.004)

Table 1: Measures per dialog [mean(sd)]. One turn comprises one system and one user turn.

utterances that instruct the system to move objects around the chart by referring either to the object's position or to nearby objects. 53% of these "inform" acts contain referring expressions (see lines 1 and 3 of Figure 3 for examples). In addition, it is clear that, due to the spatial nature of the tasks, the extra modality of plan images is key to successful planning, as reflected by the frequency of 'Image' dialog acts (around 16% of the total dialog acts). These DAs include the user requesting a plan image 'show_picture' or 'image_caption' where the system, either proactively or as a response to a user request, sends an image of the plan. The most used DA by the wizard was "ack" 30%, used for acknowledging information (e.g. "okay").

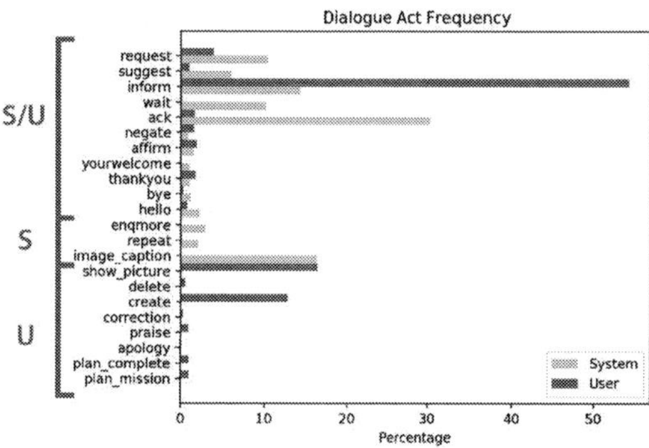

Figure 4: Dialog Act Frequency of the 22 DA types. "S" is for system only DA, "U" is for user only DA and "S/U" refers to DA that both the system and the user.

4.2 Quality of the plan

The subjective quality of the plans were measured by an expert, who has worked for years on planning missions, using a 5-point scale (see Table 2). The quality of a plan was measured according to the completeness, appropriateness and if it was operationally successful. At least 45% of the plans for both missions were measured "High Quality", with a greater number of lower rated plans for Mission 1. No correlation was found between time-on-task and quality of plans, however,

subjective feedback indicates that subjects would have liked more time to improve their plans. The response time of the Wizard was slower than natural interaction (average 15sec), which is a typical issue in WoZ studies. The dataset contains plans of varying quality, which we hope will enable the system to learn better strategies for creating optimal plans, as well as, coping strategies. There is a medium-strong positive correlation of $r = 0.59$ (Spearman's Correlation) between the expertise of the subjects (as determined by the pre-questionnaire) and the quality of the plan for the first mission indicating that, perhaps unsurprisingly, the higher the expertise, the better the quality of plans.

Quality	Mission 1	Mission 2
Very High Quality	0%	9%
High Quality	45%	45%
Neutral	9%	36%
Low Quality	18%	9%
Very Low Quality	27%	0%

Table 2: The quality of all 22 plans measured by an expert using a 5-point Likert scale.

4.3 Subjective Measures

The post-task questionnaire measured the subjective scores for User Satisfaction (US), the pace of the experiment and the importance of multimodality. Specifically the following questions were asked on a 5-point Likert Scale:

- Q1: I felt that VERSO understood me well
- Q2: I felt VERSO was easy to understand
- Q3: I knew what I could say at each point in the interaction
- Q4: The pace of interaction of VERSO was appropriate
- Q5: VERSO behaved as expected
- Q6: It was easy to create a plan with VERSO

- Q7: From my current experience with using VERSO, I would use the system regularly to create plans
- Q8: The system was sluggish and slow to respond (reversed)
- Q9: The screen shots of the plan were useful
- Q10: The screen shots of the plan were sent frequently.

Mean US is 3.5 out of 5, calculated as an average of Q1-7, which are questions adapted from the PARADISE evaluation framework (Walker et al., 1997). Q8 reflects the speed of the interaction with the mean/mode/median as 4/4/4. This score is reversed and so these high scores indicate high perceived slowness. As mentioned above, this is a common problem with wizarding set-ups and will not be a problem for the final implemented system. Q9 and Q10 refer to the images sent and we can see from the mean/mode/median of 4.6/5/5 for Q9 that images were clearly useful but perhaps could be sent more frequently (3/4/3 for Q10). The users' preference for images of plans may be related to their cognitive styles being mostly spatial.

After both tasks, we collected perceived workload using NASA Task Load Index (NASA-TLX) (Dickinson et al., 1993), where low scores indicate low cognitive workload. Our mean Raw TLX score was 46/100 ($SD = 9.08$). This mean score is comparable to a study for remote controlling robots through an interface as reported in (Kiselev and Loutfi, 2012). Further analysis and data collection would be needed to understand the user workload with respect to interaction phenomena observed in the corpus.

4.4 Qualitative Feedback

Subjects were asked two open questions of what they liked or not about VERSO. An inductive, thematic analysis was done using grounded theory with open coding (Strauss, 1987). Themes identified include:

Theme 1 Suggestions for extra functionality: Due to delays some subjects were not sure if the program crashed. We had a dialog act "wait" but feedback indicated it would be better to have a visual indicator as well. Note, in the actual future working system, we will not have the same delays as in the WoZ experiment.

Theme 2 Chart meta-data: Some subjects (P5 most specifically) desired more meta-data on the plan images they were receiving when referring to an object. When performing spatial tasks on the chart, clear referring expressions are crucial and meta-data on the chart, such as entity names (as with the Map Task (Anderson et al., 1991) landmarks), would help establish grounded referring expressions.

In our case, some of the referring expressions were names decided on between the Wizard and the subject, e.g. survey3. However, if the subject uses such objects as points of reference, e.g. "place target1 near survey3", this can become problematic when the object ("survey3") could measure up to a mile width because the exact location for "target1" is ambiguous.

Theme 3 Mixed initiative & Handling multiple requests: The WoZ interface was designed as a mixed-initiative dialog system, capable of suggesting actions and the subjects seem to like this type of interaction. Also noted was the 'system's' ability to handle multiple requests in a single utterance, which will need to be implemented in the final system.

5 Discussion and Future work

This paper presents a two-wizard WoZ study for collecting data on a collaborative task, identifying the importance of mixed modalities and object referencing, for successful interaction during mission planning. Further data collection on Amazon Mechanical Turk using Open Street Maps will be conducted in order to reach a wider audience and compensate for the gender imbalance.

Deep learning methods have surpassed human performance in a variety of tasks and one crucial factor for this achievement is the amount of data used to tune these models. However, to be able to learn from limited amounts of data will be key in moving forward (Daugherty and Wilson, 2018).

In future work, the corpus described here will be used in the development of a mixed-initiative data-driven multimodal conversational agent, for planning missions collaboratively with a human operator. With the collected WoZ data, we can capture the main strategies of how to plan a mission and make data-driven simulations possible. Therefore, we can train a Reinforcement Learning agent on simulated dialogs that are fully data-driven with the reward function being derived from our subjects' preferences, optimizing for plan quality and speed. Moreover, supervised approaches that require less data to learn, such as the Hybrid Code Networks (HCN) (Williams et al., 2017), could be used for the creation of such a system. Finally, the system will be compared to a baseline in a further human evaluation study.

References

Anne H. Anderson, Miles Bader, Ellen Gurman Bard, Elizabeth Boyle, Gwyneth Doherty, Simon Garrod, Stephen Isard, Jacqueline Kowtko, Jan McAllister, Jim Miller, Catherine Sotillo, Henry S. Thompson, and Regina Weinert. 1991. The HCRC Map Task Corpus. *Language and Speech*, 34(4):351–366.

A. Das, S. Kottur, K. Gupta, A. Singh, D. Yadav, J. M. F. Moura, D. Parikh, and D. Batra. 2017. Visual dialog. In *Proceedings of 2017 IEEE Conference on Computer Vision and Pattern Recognition (CVPR)*, pages 1080–1089.

Paul R. Daugherty and H. James Wilson. 2018. *Human + machine : reimagining work in the age of AI / Paul R. Daugherty, H. James Wilson*. Harvard Business Review Press Boston, Massachusetts.

John Dickinson, Winston D. Byblow, and L.A. Ryan. 1993. Order effects and the weighting process in workload assessment. *Applied Ergonomics*, 24(5):357 – 361.

Rui Fang, Changsong Liu, Lanbo She, and Joyce Y. Chai. 2013. Towards situated dialogue: Revisiting referring expression generation. In *Proceedings of the 51st Annual Meeting of the Association for Computational Linguistics*, pages 392–402. Association for Computational Linguistics (ACL).

Brent Harrison, Upol Ehsan, and Mark O. Riedl. 2018. Rationalization: A neural machine translation approach to generating natural language explanations. In *Proceedings of the 2018 Conference on Artificial Intelligence, Ethics and Society*.

Andrey Kiselev and Amy Loutfi. 2012. Using a mental workload index as a measure of usability of a user interface for social robotic telepresence. *2nd Workshop of Social Robotic Telepresence in Conjunction with IEEE International Symposium on Robot and Human Interactive Communication 2012*.

Nikita Kitaev, Jin-Hwa Kim, Xinlei Chen, Marcus Rohrbach, Yuandong Tian, Dhruv Batra, and Devi Parikh. 2019. Codraw: Collaborative drawing as a testbed for grounded goal-driven communication. *International Conference on Learning Representations*.

David Lane, Keith Brown, Yvan Petillot, Emilio Miguelanez, and Pedro Patron. 2013. *An Ontology-Based Approach to Fault Tolerant Mission Execution for Autonomous Platforms*, pages 225–255. Springer New York, New York, NY.

Jiwei Li, Michel Galley, Chris Brockett, Georgios Spithourakis, Jianfeng Gao, and Bill Dolan. 2016. A persona-based neural conversation model. In *Proceedings of the 54th Annual Meeting of the Association for Computational Linguistics*, pages 994–1003. Association for Computational Linguistics.

Emilio Miguelanez, Pedro Patron, Keith E Brown, Yvan R Petillot, and David M Lane. 2011. Semantic knowledge-based framework to improve the situation awareness of autonomous underwater vehicles. *IEEE Transactions on Knowledge and Data Engineering*, 23(5):759–773.

Dipendra Misra, Andrew Bennett, Valts Blukis, Eyvind Niklasson, Max Shatkhin, and Yoav Artzi. 2018. Mapping instructions to actions in 3d environments with visual goal prediction. In *Proceedings of the 2018 Conference on Empirical Methods in Natural Language Processing*, pages 2667–2678, Brussels, Belgium. Association for Computational Linguistics.

Yvan Petillot, Chris Sotzing, Pedro Patron, David Lane, and Joel Cartright. 2009. Multiple system collaborative planning and sensing for autonomous platforms with shared and distributed situational awareness. In *Proceedings of the AUVSI's Unmanned Systems Europe, La Spezia, Italy*.

Verena Rieser. 2008. *Bootstrapping Reinforcement Learning-based Dialogue Strategies from Wizard-of-Oz data*. Ph.D. thesis, Saarland University, Saarbruecken Dissertations in Computational Linguistics and Language Technology.

David Schlangen. 2016. Grounding, Justification, Adaptation: Towards Machines That Mean What They Say. In *Proceedings of the 20th Workshop on the Semantics and Pragmatics of Dialogue (JerSem)*.

Anselm L Strauss. 1987. *Qualitative analysis for social scientists*. Cambridge University Press.

Marilyn A. Walker, Diane J. Litman, Candace A. Kamm, and Alicia Abella. 1997. Paradise: A framework for evaluating spoken dialogue agents. In *Proceedings of the 36th Annual Meeting of the Association for Computational Linguistics and 17th International Conference on Computational Linguistics, ACL '98*, pages 271–280, Stroudsburg, PA, USA. Association for Computational Linguistics.

Tsung-Hsien Wen, David Vandyke, Nikola Mrkšić, Milica Gasic, Lina M. Rojas Barahona, Pei-Hao Su, Stefan Ultes, and Steve Young. 2017. A network-based end-to-end trainable task-oriented dialogue system. In *Proceedings of the 15th Conference of the European Chapter of the Association for Computational Linguistics*, pages 438–449. Association for Computational Linguistics.

Jason D. Williams, Kavosh Asadi, and Geoffrey Zweig. 2017. Hybrid code networks: practical and efficient end-to-end dialog control with supervised and reinforcement learning. In *Proceedings of the 55th Annual Meeting of the Association for Computational Linguistics*, pages 665–677, Vancouver, Canada. Association for Computational Linguistics.

¿Es un plátano? Exploring the Application of a Physically Grounded Language Acquisition System to Spanish

Caroline Kery   **Francis Ferraro**   **Cynthia Matuszek**

University of Maryland Baltimore County, Baltimore, Maryland

ckery1@umbc.edu   ferraro@umbc.edu   cmat@umbc.edu

Abstract

In this paper we describe a multilingual grounded language learning system adapted from an English-only system. This system learns the meaning of words used in crowd-sourced descriptions by grounding them in the physical representations of the objects they are describing. Our work presents a framework to compare the performance of the system when applied to a new language and to identify modifications necessary to attain equal performance, with the goal of enhancing the ability of robots to learn language from a more diverse range of people. We then demonstrate this system with Spanish, through first analyzing the performance of translated Spanish, and then extending this analysis to a new corpus of crowd-sourced Spanish language data. We find that with small modifications, the system is able to learn color, object, and shape words with comparable performance between languages.

1 Introduction

With widespread use of products like Roombas, Alexa, and drones, robots are becoming commonplace in the homes of people. We can see a future where robots are integrated into homes to provide assistance in many ways. This could be especially beneficial to elders and people with disabilities, where having someone to help with basic tasks could be what allows them to live independently (Broekens et al., 2009). Natural language is an intuitive way for human users to communicate with robotic assistants (Matuszek et al., 2012a). Grounded Language Acquisition is the concept of learning language by tying natural language inputs to concrete things one can perceive. This field of study looks to train language and perceptual skills simultaneously in order to gain a better understanding of both (Mooney, 2008). Work in this field is critical for building robots that can learn about their environments from the people around them.

For such a system to truly be useful for the average user, it is not enough to merely train a robot how to recognize everyday objects and actions in a lab. Much like toddlers who grow up in a family and surrounding culture, a service robot should be ideally able to learn the acronyms, nicknames, and other informal language that happens naturally in human interaction. It logically follows that a truly well-designed system should not only be able to handle vocabulary differences between users but also users that speak different languages. There are thousands of official languages spoken around the world, and many more dialects. In the United States alone, around 21 percent of residents speak a non-English language as their primary language at home (United States Census Bureau, US Department of Commerce, 2017). Grounded Language Acquisition takes many of its roots from Natural Language Processing, which in the past has had an unfortunate habit of focusing on English-centric methods. This often leads to systems that perform very well in English and "well enough" in other languages.

In this paper, we take an existing grounded language acquisition system (Matuszek et al., 2012b; Pillai and Matuszek, 2018) designed for grounding English language data and examine what adaptations are necessary for it to perform equally well for Spanish. We explore the extent to which machine translated data can assist in identifying linguistic differences that can impact system performance. We then collect a new comparable corpora of crowd-sourced Spanish language data and evaluate it on the system with and without our proposed modifications.

Proceedings of SpLU-RoboNLP, pages 7–17

Minneapolis, MN, June 6, 2019. ©2019 Association for Computational Linguistics

2 Previous Work

In this section, we describe relevant previous works in grounded language acquisition and multilingual natural language processing. While there has been past work to apply grounded language learning systems to multiple languages (Chen et al., 2010; Alomari et al., 2017) to our knowledge there have been few efforts in the space of non-English grounded language learning where comprehensive analysis was done to diagnose differences in performance between languages and work to mitigate these differences.

2.1 Grounded Language Acquisition

Language grounding can be done in many ways. There is a significant community within computer vision that works on object recognition with the help of captions (Krishna et al., 2017; Gao et al., 2015). These efforts ground objects found in images with words and relations stated in the captions. A multilingual example of this by Gella et al. (2017), used images as pivots between English and German image descriptions. This paper has a similar task of mapping language to images, but does so on a token level, and does not attempt to combine information between the Spanish and English corpora. In addition, the image data we are using includes depth information, as we are simulating the visual percepts of a robot. It must be noted that this differs from other works that use additional products of robotic percepts like video data, eye tracking, and other forms of gesture recognition (Chen et al., 2010; Alomari et al., 2017; Kollar et al., 2014; Yu and Ballard, 2004). In the robotics space, many works tie language grounding to enable actions like pathfinding (Matuszek et al., 2012a), household tasks (Alomari et al., 2017), and building (Brawer et al., 2018). While performing practical tasks is the eventual goal of our grounded language system, the current system focuses on the first step: building representations of objects and how they are described (nouns and adjectives).

There are a few examples of language grounding in multiple languages (Chen et al., 2010; Alomari et al., 2017). Several works tested their system in a language besides English and presented the results for both. While this showed that their systems could handle multiple languages, none provided an in-depth analysis into the differences in performance for their systems, or extrapolated

past the two languages. Our work seeks to examine and identify causes of differences in performance. While our current work only displays this system with Spanish, we plan to extend our framework to additional languages in the near future.

2.2 Multilingual Natural Language Processing

There is a strong multilingual community in the broader field of NLP working in many different aspects, such as machine translation (Wu et al., 2016) or multilingual question answering (Gao et al., 2015). Some works dive deep into specific language pairs to evaluate how differences between the languages complicate translation (Alasmari et al., 2016; Gupta and Shrivastava, 2016; Ghasemi and Hashemian, 2016). Several work with Spanish and English specifically (Le Roux et al., 2012; Pla and Hurtado, 2014). Analyses such as these helped to shape our analysis when comparing the English and Spanish data performance, and enabled us to predict points where linguistic differences could impact performance.

There are quite a few examples in literature of taking a system designed for English and adapting it for multilingual use (Daiber et al., 2013; Gamon et al., 1997; Macdonald et al., 2005; Poesio et al., 2010; Jauhar et al., 2018). Sometimes this involves manually recreating aspects of the system to match the rules of the other language (Gamon et al., 1997), or utilizing parallel corpora to transfer learning between languages (Jauhar et al., 2018). Other projects look to make an English system "language agnostic" (not biased towards any one language) by editing parts of the preprocessing algorithm (Daiber et al., 2013; Wehrmann et al., 2017). The first method introduces a lot of additional complications such as manual rule-building, so it may seem attractive to make a system that is completely language-blind. The problem with this is that even generalized preprocessing techniques are often still biased towards languages with English-like structures (Bender, 2009), and in avoiding specifying anything about the language one can miss out on common properties within language families that could increase performance. For this paper, we strive to find common ground between making our system as generalized as possible and taking specific linguistic structures into account if necessary.

One significant difference between our research

and many works in grounded language acquisition is that our system is entirely trained off of noisy short descriptions collected without filtering. This has very different characteristics from the more common corpora built off of newswire and other forms of well-written text (a very common one is multilingual Wikipedia), or data that has been placed into structures like trees (Le Roux et al., 2012). Our data is prone to errors in grammar and misspellings; in this regard, our data is most like that of works that use Twitter data (Pla and Hurtado, 2014). However, in contrast to (Pla and Hurtado, 2014), our system only uses token extraction to find the relevant images to extract features from, rather than extracting all features from just the language.

3 Approach

In this paper, instead of building a new grounded language system, we chose to start with an existing system presented by Pillai and Matuszek (2018), which we will refer to as the GLS (Grounded Language System). This system attempted to learn physical groundings of colors, shapes, and objects by tying color and depth data derived from images of various items with natural language descriptions of the images. As a broader research goal, we seek to discover how effective the GLS is at handling non-English data. We decided to start with Spanish, due to it being very similar to English. We wanted to see if and how the slight differences between the two languages would affect the relative performance of the system.

To begin our analysis we explored the performance of the system on translated Spanish data with minimal modifications. Our analysis of these results concentrated on identifying language differences between Spanish and English that introduced new complications in grounding language. We used our insights from this analysis to inform our experiments on real Spanish data collected using Amazon Mechanical Turk.

3.1 Data

Pillai and Matuszek (2018) used a Kinect depth camera to collect images of fruit, vegetables, and children's blocks of various shapes (see figure 1 for examples). There were a total of 18 object types, with four instances of each object. Each instance had around five images taken using the depth camera. For each of these images, RGB and

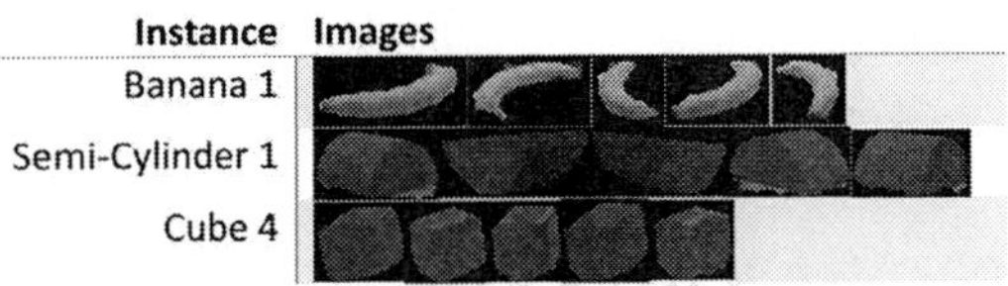

Figure 1: Examples of images of objects in the original dataset. Each object had several examples called "instances" and images of each instance were taken from several angles.

HMP-extracted kernel descriptors (Bo et al., 2011; Lai et al., 2013) were extracted from the RGB-D data. The authors then collected descriptions of these images in English using Amazon Mechanical Turk. About 85 descriptions were collected for each instance, for a total corpus of about six thousand descriptions. As we discuss in section 6, our own data collection process replicated this setup.

3.2 Grounding System

The GLS learned classifiers for meaningful tokens in the categories of color, shape, and object in an unsupervised setting. The system used the Mechanical Turk descriptions to identify which images were positive or negative examples of each token. Images that were described with a particular token often were assumed to be examples of that token. To find negative examples of tokens, the GLS used document similarity metrics to compare the descriptions for instances in vector space. The instances that were sufficiently far away in vector space from the identified positive instances for a token that had also never been described with that token were then chosen as negative examples of that token. For example, suppose the system were finding positive and negative instances for the token "carrot." A positive instance identified might be "carrot 4." In the document vector space, the instances with the descriptions most different from "carrot 4" would be "arch 1" and "cuboid 4," while instances like "tomato 2" and "cucumber 3" are closer but still different enough to possibly qualify as negative examples of the token "carrot."

Tokens that did not have any negative examples or had fewer than three positive examples were thrown away, with the assumption that there was not enough data to learn a classifier. The final classifiers were scored using the downstream task of image classification. Held-out positive and negative examples were presented, and the classifiers were judged by how well they could identify which examples were positive or negative.

3.3 Our Modifications

Our research focused on taking the existing system and expanding it to work with Spanish. In the immediate sense, there were low level changes that had to be made throughout the code. English uses very few accents and many of the files had to have their encoding specified as unicode to handle non-ASCII characters. These changes, though minor, reflect a potential barrier to the application of research in new settings.

In addition to these minor fixes, more substantial changes had to be made to the system that preprocessed the image descriptions. The original GLS used an English WordNet lemmatizer. Lemmatizers are tools that take conjugated words like "walking" or "brushes" and attempt to turn them into un-conjugated versions like "walk" or "brush." This step can be very helpful for making sure different versions of the same word are conflated. While this system worked well for English tokens, non-English lemmatizers proved difficult to find. Since we would ideally like our adaptations to the system to generalize well to other future languages, we decided to first remove the lemmatization step entirely, and later when this proved unsatisfactory for Spanish (see Sect. 5), we replaced the lemmatization step with a more available but rougher stemming step. Stemmers also attempt to remove conjugations from words, but they typically do so by chopping off common affixes without attempting to end up with a real word at the end. Words like "eating" will become "eat," but words like "orange" may become "orang."

Another step that we modified was the removal of "stop words." In the original system non-meaningful words like "the," "and," "or," and "if" were removed from the English data using a list of predefined words. This was an important step as it ensured that the system did not attempt to learn groundings for words like "and." At the same time, we found that there were a number of words like "object," "picture," or "color" that were used so often in the descriptions that they held little physical meaning. These are designated as "domain-specific stop words," which refer to words that in general cases hold meaning, but for the particular domain have been rendered meaningless by their frequent and varied use. We found that these words could be identified by their inverse-document-frequency (IDF), where each "document" is the concatenation of all descriptions for an instance.

4 Analysis with Translated Data

For our preliminary experiments, we only had access to the English corpora from Pillai and Matuszek (2018). We wanted to get baselines in how a Spanish corpora might perform. To do this, we translated the existing English phrases to Spanish through Google Translate's API (Wu et al., 2016).

4.1 Translation Accuracy

As a sanity check on the quality of translation, the translated text was translated back into English (once again with Google translate's API) and the English and back-translated English phrases were compared manually to see if their overall meanings were preserved. A total of 2,487 out of the 6,120 (around 40%) phrases remained exactly the same between translations. For the remaining 60%, five hundred back-translated phrases were randomly selected and manually compared to their original English version (see table 3 for examples). Approximately 87% of the phrases examined preserved their meaning between translations, so we estimated from this that about 90% of the phrases were translated accurately (shown in figure 2).

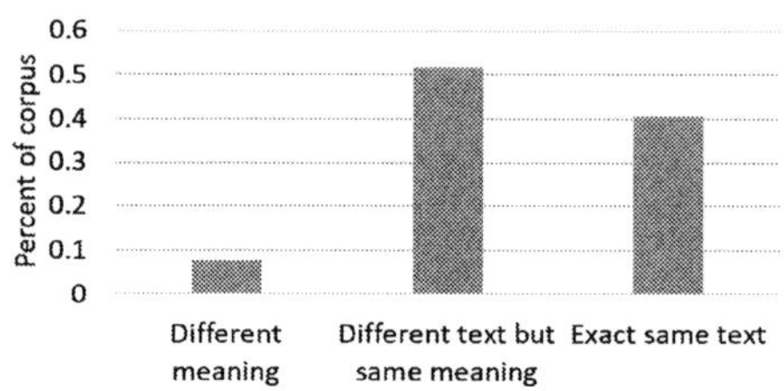

Figure 2: Breakdown of meaning preservation for English and English-Spanish-English translation.

For those phrases that did not translate accurately back to English, we observed a number of patterns. Some of them were simply due to ambiguities with the meaning of a word where the wrong one was selected during one of the translations (as an example, for the bottom row of table 3, "forma" can mean "shape" or "way"). A common example of this was the phrase "this is a red cabbage" becoming "this is a red wire," which happened six times out of the five hundred selected phrases. Another error that occurred three times was "laying on its side" becoming "set

Image ID	Original English	Spanish Translation (Google API)	Back-translated English (Google API	Same Meaning?
Orange 2	This fruit is called an orange	Esta fruta se llama naranja	This fruit is called orange	Yes
Cuboid 4	This is a picture of rectangular shaped blue coloured solid block	Esta es una foto de bloque sólido de color azul con forma rectangular.	This is a solid block photo of blue with rectangular shape.	No
Lime 2	It is a lime	Es una lima	It's a lime	Yes
Cuboid 2	This is a block The block is green The background is black The green block is laying on its side	Esto es un bloque El bloque es verde El fondo es negro El bloque verde está de lado	This is a block The block is green The background is black The green block is on its side	Yes
Cuboid 3	THIS IS A SHAPE	Esto es una forma	This is a way	No

Figure 3: Samples of English descriptions that were translated into Spanish and then back into English. The column on the right indicates if the meaning of the original English text matches the final back-translated English

aside," since the Spanish phrase "puesta de lado" can mean "put sideways" but also "set aside."

Other translation errors could be related to differences in Spanish and English structures. The pronoun "it" commonly became "he," as Spanish nouns are gendered. Phrases with many adjectives saw them switching places with each other and the nouns they were attributed to. For example, "This is a picture of rectangular shaped blue colored solid block" became "This is a solid block photo of blue with rectangular shape." This confusion could be due to differences in the rules of adjective ordering between English and Spanish.

5 Scores for English and Google Translated Spanish

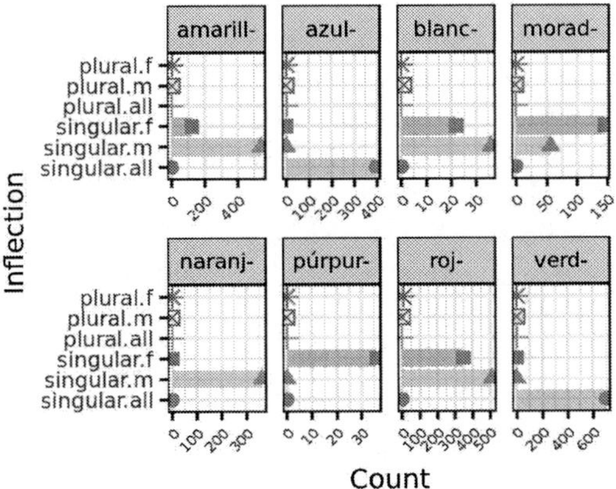

Figure 4: Proportion of color word forms in raw translated Spanish.

For the first experiment, we trained the model on the translated Spanish and English corpora with minimal preprocessing (lowercasing and removing punctuation), and tested the color tokens only. Our goal was to get a baseline for how the system would perform using words that would be easy to compare between languages. It was expected that the Spanish corpus would perform worse, since it was not perfectly translated. When the tests were run, the translated Spanish did perform slightly worse (see figure 5), but an additional interesting issue emerged.

Spanish is a highly inflected language (Le Roux et al., 2012) and unlike English has adjective-noun agreement. This means that a simple color word like "red," could translate to "rojo," "roja," "rojos," or "rojas" depending on the gender and plurality of the noun it is describing. For the learning system this meant that the possible positive instances for color words could be split between the various forms, since different descriptions of the same object might use a different form depending on the structure of the sentence. We can see from figure 4, that in the overall translated corpus, the color words were split between different conjugations. This led to the hypothesis that some form of lemmatization or stemming would be necessary for Spanish, in a way that would have been less essential for English.

We processed both the translated Spanish and English descriptions with a Snowball stemmer (Porter, 2001). We chose this stemmer as it is readily available for a wide variety of languages through the nltk library. See results in Fig. 5.

We can see from figure 5 that applying stemming to the translated Spanish descriptions had a small positive effect on the F1-scores of the color

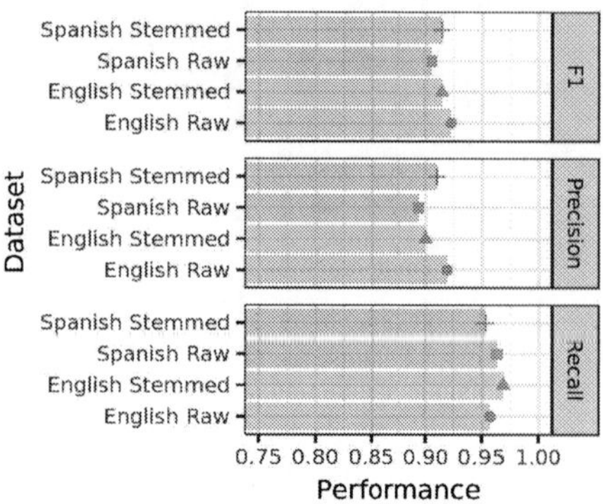

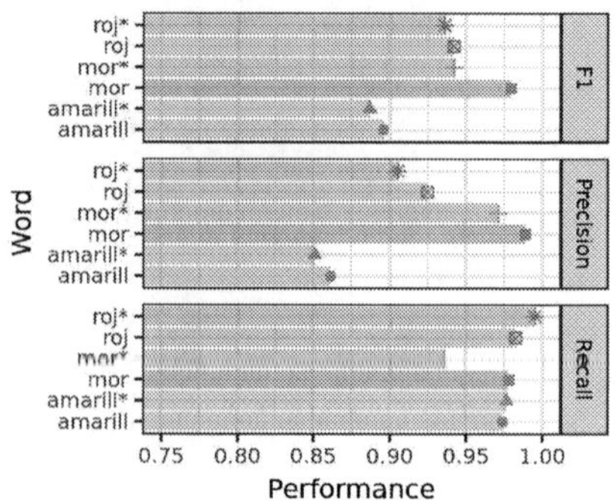

Figure 5: Average Scores for English and Google Translated Spanish.

Figure 7: Comparison between the average of scores for various conjugations of color words (shown as *) and the scores of the stemmed versions.

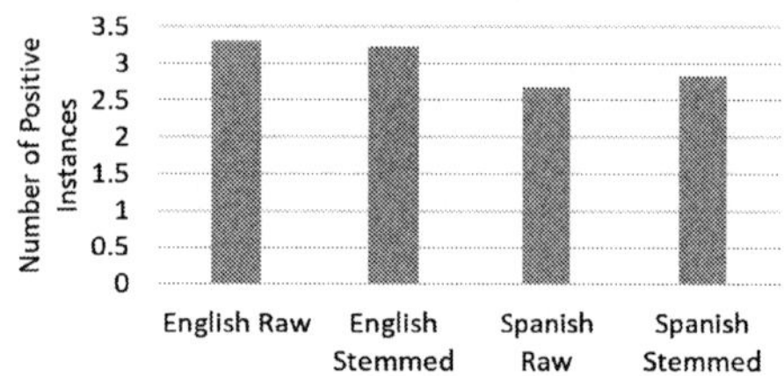

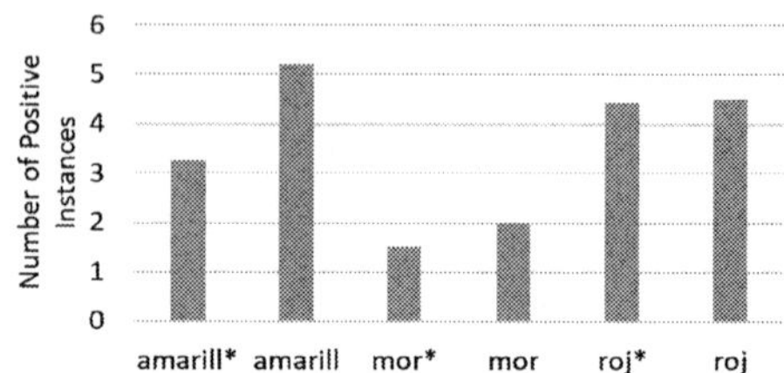

Figure 6: Average number of positive instances for English and Google Translated Spanish, stemmed and un-stemmed.

Figure 8: Comparison between the average number of positive instances across color word conjugations (see figure 6) and the number of positive instances of their stemmed forms.

classifiers. It also slightly raised the average number of positive instances per token, since stemming allowed instances that were split between small counts of several forms of a word to see them as the same word. We can see this in more detail in figures 7 and 8, which show the difference between the average of the scores for the various forms of color words in the unstemmed data (for example amarilla and amarillo would be averaged as amarill*), and the stemmed score of the stemmed form.

We can see in figure 7 that for the three colors shown, stemming always increased the average precision for that color, but could reduce recall. In addition from figure 8, we see that some of the colors had a large increase in average positive instances, while others did not. This was likely due to a case where many instances labeled with "rojo" also saw enough "roja" that it was a positive instance for both. When looking at the counts per instance, we found that for the 23 instances that had the token "roj" in their stemmed descriptions, 16 were positive examples of both "roja" and "rojo" in the un-stemmed version. For objects like cabbages (coles) and plums (ciruelas), "roja"

was used dramatically more, while for tomatoes (tomates), cubes (cubos), and cylinders (cilindros) "rojo" appeared more.

As a final check, we examined the number of occurrences over all descriptions of each instance of the stemmed and un-stemmed versions of color words. For most of the colors, instances were often split between possible conjugations. For "amarill" (yellow), there were five instances where the individual counts of both un-stemmed forms of yellow: "amarillo" and "amarilla" were less than the threshold for a positive instance, while the stemmed version "amarill" was able to overcome that threshold. This is shown in the more dramatic increase in number of positive examples in figure 8. The effect on the scores is more complicated, since very yellow instances often had 50 or more occurrences of "amarill." Because of the inherent messy nature of the data, instances with low but still significant counts of tokens (more than five occurrences) were much more likely to be falsely positive examples that could damage a classifier. We see this in figure 9 where the instance "eggplant 1" was called green seven times in the En-

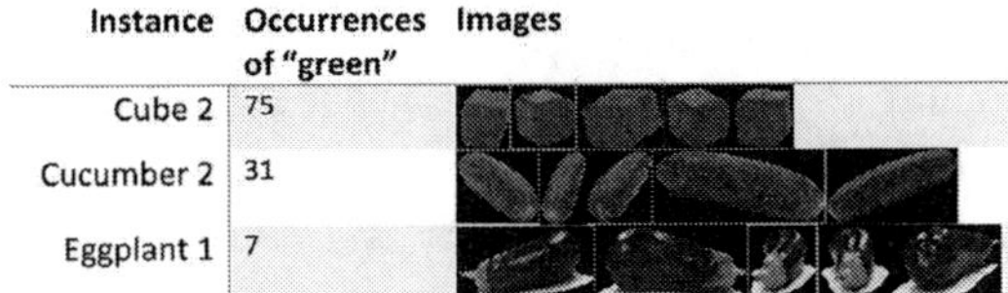

Instance	Occurrences of "green"	Images
Cube 2	75	
Cucumber 2	31	
Eggplant 1	7	

Figure 9: Sample of instances that had more than five occurrences of "green" in the English corpora.

glish data. This is clearly because the stem of the fruit is green. However, a simple classifier may be confused by this instance, since it is mostly purple.

6 Collection of Real Spanish Data

Exploring comparisons between English and translated Spanish enabled us to get a basic idea of how Spanish descriptions might differ from English. However, in order to truly compare the languages, we needed to collect real Spanish data. We attempted to follow the methods described by Pillai and Matuszek (2018) as closely as possible to obtain comparable Spanish data to their English data. We utilized Amazon Mechanical Turk to collect Spanish descriptions of the images in the database.[1] In addition, workers were required to have at least fifty HITs accepted before being eligible to work on our HITs. To avoid biasing the workers towards a particular type of description, we provided no example descriptions.

We excluded data from a small number of workers who did not follow the directions (for example, responding in English or randomly selecting adjectives) and obtained additional high quality data to replace their submissions. All other submissions were accepted. This allowed for a wide variety of answers. One worker might simply name a carrot, while another would describe how it tastes, what foods it goes well in, or where it comes from. The English dataset was similarly noisy. This is desirable, as a robot that is trying to learn language from an average user must be able to handle the many ways in which a user might choose to introduce a new object.

One possible danger in collecting Spanish data that we considered was that someone might be responding in English and using a translation tool. We attempted to check for this by comparing our real Spanish data to the translated Spanish data. We found that short descriptions like "Esto es un limón" (this is a lemon) had a large amount of

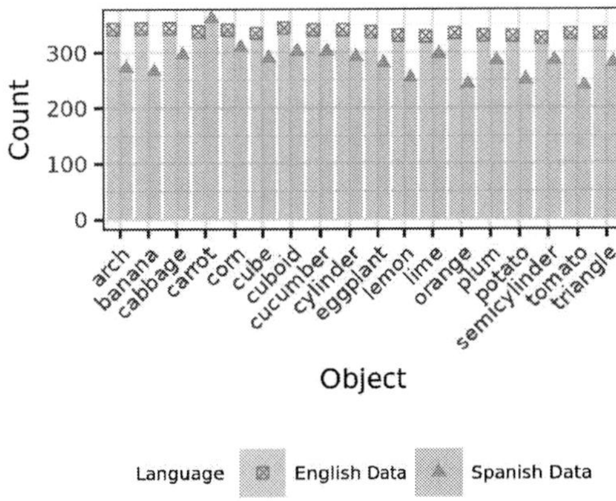

Figure 10: Total number of descriptions collected per object in Spanish and English.

overlap, but in general most of the Spanish descriptions were longer and did not mirror any of the translated results. In future work, we hope to find a better method to control for respondents who don't actually speak the language, likely by requiring the completion of some short preliminary task like text summarization or more complex image captioning.

The total number of Spanish descriptions per object type was on average slightly lower than in the English corpus (see figure 10). We controlled for this in the results (section 7) by taking several random subsets of both corpora such that each instance had an equal number of Spanish and English descriptions and averaging the results.

7 Comparison of Spanish and English

7.1 Overall Scores

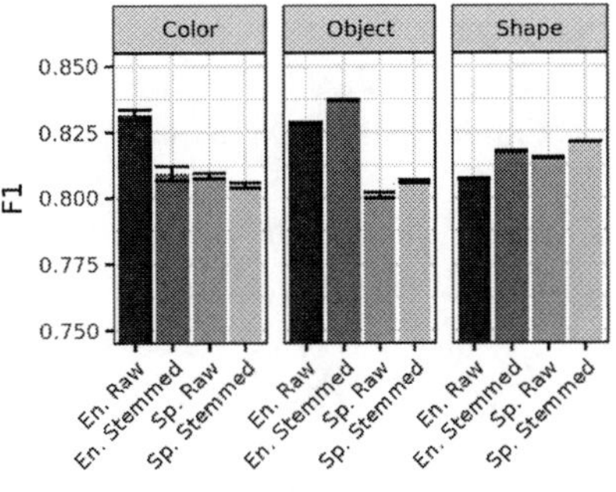

Figure 11: Average F1 scores for English and Spanish classifiers, stemmed and un-stemmed, for each classifier type. The error bars show the variance of these scores across all runs, which was fairly low.

---

[1]This was accepted as an IRB exempt study.

In figure 11, we see the final averaged F1-Score for the color, shape, and object classifiers between the original English and the collected Spanish descriptions. Each score was found by averaging the results of twenty evaluation runs each of ten train-test splits. These scores were averaged across all tokens learned, without specifically sub-setting for the tokens that naturally represented colors, shapes, or objects. In general, the scores were fairly similar, varying between 0.8 and 0.84. From the small differences we see that stemming appeared to benefit the Spanish data for learning object and shape classifiers, but slightly hurt the performance for color classifiers. Un-stemmed English performed better than either Spanish version for color and object classifiers. Much like with Spanish, stemming appeared to help the shape and object classifiers, and hurt the color ones.

7.2 The Effect of Stemming

As one can see from figure 11, the effect of Stemming on the F1-Scores of the English and Spanish classifiers was not consistent. For both the object and shape classifiers, stemming appeared to either benefit or have little impact on the object recognition task. For the color tokens, stemming either barely impacted or lowered the scores.

Stemming can cause words to be conflated correctly or incorrectly. Incorrect stemming can certainly cause problems, where tokens are conflated that shouldn't be (Porter, 2001), or words that should be conflated are not. However, as discussed earlier, it is also possible for correct stemming to cause an instance to barely meet the threshold for being a positive example of a particular token 9, when perhaps that instance is not a good example of that token in reality. This was a particularly likely occurrence due to the inherent messiness of the data and the fact that the GLS based the classification label off of these messy descriptions. Due to this, and the high amount of conjugation in Spanish, it was decided that stemming likely would not negatively impact the learning system, and should most likely be employed.

7.3 Accents

One interesting difference that stood out when examining the real Spanish data was the use of accents. Unlike with the translated data, the real Spanish data was inconsistent with its usage of accents. While a majority of workers used accents where they were supposed to go, a not-insignificant percentage of them left them out (see figure 12 for examples). This is likely because those workers did not have easy access to a keyboard with accented characters, and thus chose to leave them off. We can see in figure 12 that for common accented words, this had the effect of splitting the data. Luckily, the snowball stemmer (Porter, 2001) automatically removed these accents. We can see in figure 12 that after stemming, the counts for the accented and unaccented versions of the token were combined. The combined classifier did not always have a higher score on the testing data, for similar reasons to those discussed in section 7.2.

7.4 Stopwords

Without employing stop word removing during preprocessing, the system learned a total of ten words that could be classified as general stop words for English and eight for Spanish (see figure 13). This means that for these words, there was at least one instance where the word did not appear in any description. For Spanish, the tokens "de," "es," "una," "y," and "se," and for English the tokens "this," "is," and "a" all had zero negative instances and were appropriately removed.

Figure 13 also shows tokens that appeared in the bottom 2% of tokens when sorted by IDF score. This was our way of estimating "domain-specific stop words." Note that there were quite a few nltk stop words that also had very low IDF scores. The IDF method identified tokens like "object", or "looks" which were used very often in the English descriptions and had little physical meaning. Figures 14 and 15 show how removing each type of stop word impacted the scores of the raw classifiers. For both languages, the greatest impact appeared to come from removing both general purpose stop words and low-IDF tokens, though the impact was small in all cases.

For the Spanish data, the tokens "amarillo" (yellow) and "roja" (red) were included in the bottom 2% of tokens by IDF score. These were common due to the prevalence of red and yellow objects in the dataset, suggesting a more nuanced approach such as lowering the threshold for the percent of low-IDF tokens to be thrown out.

8 Future Work

The work presented in this paper is ongoing. In the near future we intend to expand the analysis on

	Token	Count	F1-Score	Token	Count	F1-Score	Token	Count	F1-Score
English	corn	261	0.926508	banana	261	0.82946	lemon	252	0.907535
Spanish (accented)	maíz	117	0.802675	plátano	90	0.65640	limón	165	0.898421
Spanish (no accent)	maiz	65	0.793374	platano	47	0.66132	limon	118	0.866587
Spanish stemmed	maiz	182	0.835578	platan	140	0.65773	limon	283	0.840359

Figure 12: Object Scores for three Spanish that could be written with and without accents. Note that stemming removed accents, conflating stemmed and un-stemmed versions together.

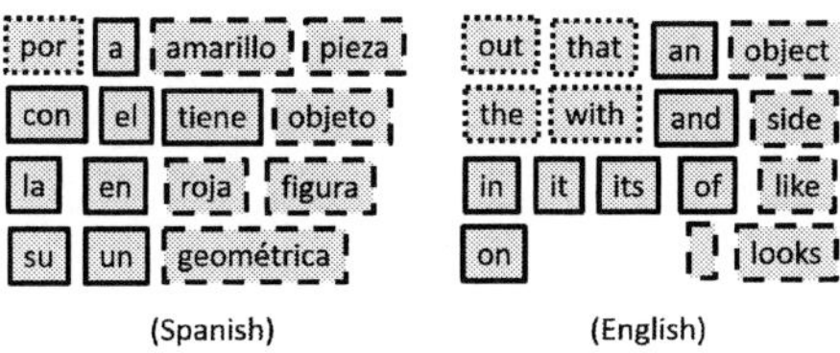

Figure 13: Stop words that appeared often enough to have classifiers trained on them. A dotted border indicates a stop word from the language's nltk stop word list. A dashed border indicates this token was in the top 2% tokens by ascending IDF score. A solid border means the token appeared in both lists.

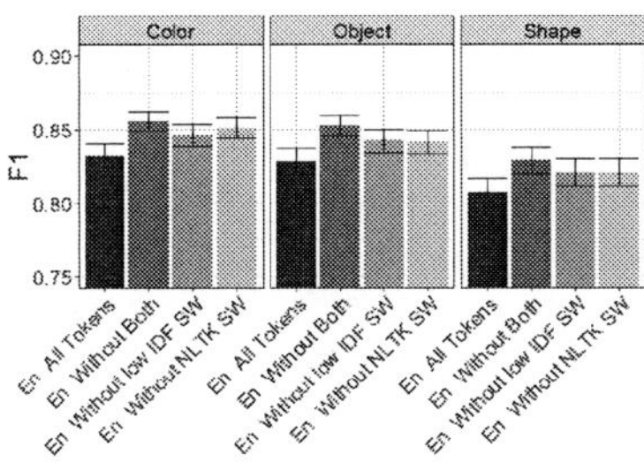

Figure 14: The impact on the average F1-score of removing nltk stop words versus removing the lowest 2% tokens by IDF score for English.

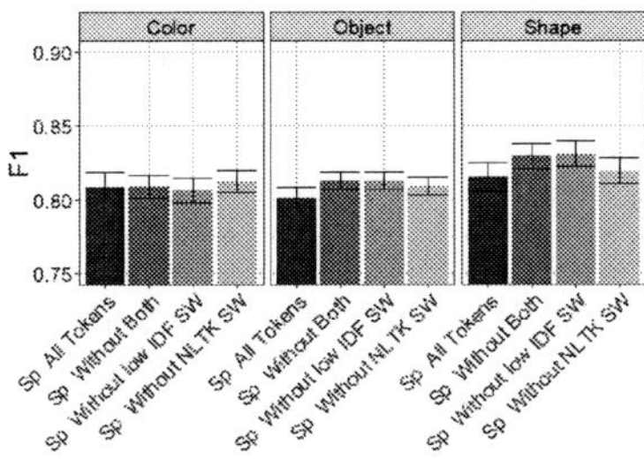

Figure 15: The impact on the average F1-score of removing nltk stop words versus removing the lowest 2% tokens by IDF score for Spanish.

the Spanish data. In addition many other possible techniques like spell-checking or synonym identification could be used to improve the ability of the system to handle the messy data.

A major next step for this research is to run our analysis on a language that is very different from English. For this, we intend to look next at Hindi. Hindi is the native language for hundreds of millions of people (India: Office of the Registrar General & Census Commissioner, 2015). It is from a different language family than English or Spanish, has a wide variety of dialects with small linguistic differences, and uses its own script. We anticipate that these properties will make Hindi a complicated and interesting language to analyze, and that doing so will introduce many new considerations for the grounded language system.

9   Conclusion

We have proposed adaptations to expand an existing unsupervised grounded language acquisition system (Pillai and Matuszek, 2018) to work with Spanish data. We discussed our initial observations with Google translated Spanish, and explored the extent to which these observations could be extended to real Spanish data collected through Amazon Mechanical Turk. Through our experiments, we were able to identify several differences between the two languages that had to be addressed in the system to attain comparable results. At the same time, we did not find that Spanish did significantly worse than English even before applying additional steps. In general, the existing system with slight modifications seems to work fairly well for both languages, which is promising when considering its applicability to real-life situations.

References

Jawharah Alasmari, J Watson, and ES Atwell. 2016. A comparative analysis between arabic and english of the verbal system using google translate. In *Proceedings of IMAN'2016 4th International Conference on Islamic Applications in Computer Science and Technologies*, Khartoum, Sudan.

Muhannad Alomari, Paul Duckworth, David C Hogg, and Anthony G Cohn. 2017. Natural language acquisition and grounding for embodied robotic systems. In *Thirty-First AAAI Conference on Artificial Intelligence (AAAI-17)*.

Emily M Bender. 2009. Linguistically naive!= language independent: why nlp needs linguistic typology. In *Proceedings of the EACL 2009 Workshop on the Interaction between Linguistics and Computational Linguistics: Virtuous, Vicious or Vacuous?*, pages 26–32.

Liefeng Bo, Kevin Lai, Xiaofeng Ren, and Dieter Fox. 2011. Object recognition with hierarchical kernel descriptors. In *Computer Vision and Pattern Recognition*.

Jake Brawer, Olivier Mangin, Alessandro Roncone, Sarah Widder, and Brian Scassellati. 2018. Situated human–robot collaboration: predicting intent from grounded natural language. In *2018 IEEE/RSJ International Conference on Intelligent Robots and Systems (IROS)*, pages 827–833.

Joost Broekens, Marcel Heerink, Henk Rosendal, et al. 2009. Assistive social robots in elderly care: a review. *Gerontechnology*, 8(2):94–103.

David Chen, Joohyun Kim, and Raymond Mooney. 2010. Training a multilingual sportscaster: Using perceptual context to learn language. *J. Artif. Intell. Res. (JAIR)*, 37:397–435.

Joachim Daiber, Max Jakob, Chris Hokamp, and Pablo N Mendes. 2013. Improving efficiency and accuracy in multilingual entity extraction. In *Proceedings of the 9th International Conference on Semantic Systems*, pages 121–124. ACM.

Michael Gamon, Carmen Lozano, Jessie Pinkham, and Tom Reutter. 1997. Practical experience with grammar sharing in multilingual nlp. *From Research to Commercial Applications: Making NLP Work in Practice*.

Haoyuan Gao, Junhua Mao, Jie Zhou, Zhiheng Huang, Lei Wang, and Wei Xu. 2015. Are you talking to a machine? dataset and methods for multilingual image question. In *Advances in neural information processing systems*, pages 2296–2304.

Spandana Gella, Rico Sennrich, Frank Keller, and Mirella Lapata. 2017. Image pivoting for learning multilingual multimodal representations. *arXiv preprint arXiv:1707.07601*.

Hadis Ghasemi and Mahmood Hashemian. 2016. A comparative study of" google translate" translations: An error analysis of english-to-persian and persian-to-english translations. *English Language Teaching*, 9:13–17.

Ekta Gupta and Shailendra Shrivastava. 2016. Analysis on translation quality of english to hindi online translation systems- a review. In *International Journal of Computer Applications*.

India: Office of the Registrar General & Census Commissioner. 2015. Comparative speakers' strength of scheduled languages -1971, 1981, 1991 and 2001. Archived 2007-11-30.

Sujay Kumar Jauhar, Michael Gamon, and Patrick Pantel. 2018. Neural task representations as weak supervision for model agnostic cross-lingual transfer. *arXiv preprint arXiv:1811.01115*.

Thomas Kollar, Stefanie Tellex, Deb Roy, and Nick Roy. 2014. Grounding verbs of motion in natural language commands to robots. In *Experimental Robotics. Springer Tracts in Advanced Robotics*, Springer, Berlin, Heidelberg.

Ranjay Krishna, Yuke Zhu, Oliver Groth, Justin Johnson, Kenji Hata, Joshua Kravitz, Stephanie Chen, Yannis Kalantidis, Li-Jia Li, David A Shamma, et al. 2017. Visual genome: Connecting language and vision using crowdsourced dense image annotations. *International Journal of Computer Vision*, 123:32–73.

Kevin Lai, Liefeng Bo, Xiaofeng Ren, and Dieter Fox. 2013. Rgb-d object recognition: Features, algorithms, and a large scale benchmark. In *Consumer Depth Cameras for Computer Vision: Research Topics and Applications*, pages 167–192.

Joseph Le Roux, Benoit Sagot, and Djamé Seddah. 2012. Statistical parsing of spanish and data driven lemmatization. In *ACL 2012 Joint Workshop on Statistical Parsing and Semantic Processing of Morphologically Rich Languages (SP-Sem-MRL 2012)*, pages 6–pages.

Craig Macdonald, Vassilis Plachouras, Ben He, Christina Lioma, and Iadh Ounis. 2005. University of glasgow at webclef 2005: Experiments in perfield normalisation and language specific stemming. In *Workshop of the Cross-Language Evaluation Forum for European Languages*, pages 898–907.

Cynthia Matuszek, Nicholas FitzGerald, Evan Herbst, Dieter Fox, and Luke Zettlemoyer. 2012a. Interactive learning and its role in pervasive robotics. In *ICRA Workshop on The Future of HRI*, St. Paul, MN.

Cynthia Matuszek, Nicholas FitzGerald, Luke Zettlemoyer, Liefeng Bo, and Dieter Fox. 2012b. A joint model of language and perception for grounded attribute learning. In *Proceedings of the 2012 International Conference on Machine Learning*, Edinburgh, Scotland.

Raymond Mooney. 2008. Learning to connect language and perception. In *Proceedings of the 23rd AAAI Conference on Artificial Intelligence (AAAI)*, pages 1598–1601, Chicago, IL.

Nisha Pillai and Cynthia Matuszek. 2018. Unsupervised selection of negative examples for grounded language learning. In *Proceedings of the 32nd National Conference on Artificial Intelligence (AAAI)*, New Orleans, USA.

Ferran Pla and Lluís-F Hurtado. 2014. Political tendency identification in twitter using sentiment analysis techniques. In *Proceedings of COLING 2014, the 25th international conference on computational linguistics: Technical Papers*, pages 183–192.

Massimo Poesio, Olga Uryupina, and Yannick Versley. 2010. Creating a coreference resolution system for italian. In *International Conference on Language Resources and Evaluation*, Valletta, Malta.

Martin F. Porter. 2001. Snowball: A language for stemming algorithms. *Retrieved March*, 1.

United States Census Bureau, US Department of Commerce. 2017. American community survey. Data collected from 2012-2016.

Joonatas Wehrmann, Willian Becker, Henry EL Cagnini, and Rodrigo C Barros. 2017. A character-based convolutional neural network for language-agnostic twitter sentiment analysis. In *2017 International Joint Conference on Neural Networks (IJCNN)*, pages 2384–2391. IEEE.

Yonghui Wu, Mike Schuster, Zhifeng Chen, Quoc V Le, Mohammad Norouzi, Wolfgang Macherey, Maxim Krikun, Yuan Cao, Qin Gao, Klaus Macherey, et al. 2016. Google's neural machine translation system: Bridging the gap between human and machine translation. In *CoRR*.

Chen Yu and Dana H. Ballard. 2004. A multimodal learning interface for grounding spoken language in sensory perceptions. In *ACM Transactions on Applied Perception*, pages 57–80.

Demo Paper:

From Virtual to Real: A Framework for Verbal Interaction with Robots

Eugene Joseph,
North Side Inc.
{eugene-at-northsideinc-dot-com}

Abstract

A Natural Language Understanding (NLU) pipeline integrated with a 3D physics-based scene is a flexible way to develop and test language-based human-robot interaction, by virtualizing people, robot hardware and the target 3D environment. Here, interaction means both controlling robots using language and conversing with them about the user's physical environment and her daily life. Such a virtual development framework was initially developed for the Bot Colony videogame launched on Steam in June 2014, and has been undergoing improvements since.

The framework is focused of developing intuitive verbal interaction with various types of robots. Key robot functions (robot vision and object recognition, path planning and obstacle avoidance, task planning and constraints, grabbing and inverse kinematics), the human participants in the interaction, and the impact of gravity and other forces on the environment are all simulated using commercial 3D tools. The framework can be used as a robotics testbed: the results of our simulations can be compared with the output of algorithms in real robots, to validate such algorithms.

A novelty of our framework is support for social interaction with robots - enabling robots to converse about people and objects in the user's environment, as well as learning about human needs and everyday life topics from their owner.

1   Background and motivation

If robots are to suitably collaborate with humans in tasks commonly occurring in settings like the home or at work, natural language interaction with robots is a must. Connell (2018, 2012) argues that the most efficient interface to universal helper robots, at least for the elderly, is direct speech.

North Side Inc. (www.northsideinc.com) originally developed a Natural Language Understanding (NLU) and Generation (NLG) pipeline (Joseph 2012, 2014) for its Bot Colony video game (www.botcolony.com). The game is based on the Bot Colony science-fiction novel (Joseph 2010), which anticipated the functionality of verbal interfaces of intelligent robots circa 2021.

An all-graphics (virtual) framework removes the constraints related to real hardware, enabling one to focus on refining an intuitive language-based human-robot interaction. With hardware out of the way, it is easier, faster and cheaper to make progress, and virtual robots are acceptable interlocutors (Bainbrige, 2008). All the functions described in the paper were implemented and can be observed in Bot Colony.

2   Requirements for natural verbal interactions with robots

When interacting verbally, a person would expect a robot to understand whatever he/she says just as well as a member of our species would (Bisk, 2016). Capabilities (i) – (v) below, implemented in our framework, are innovative features of the framework.

(i) To link language to actions and objects in the real world, one should be able to refer to objects or people using natural speech – using similar words, similar syntax, and using pronouns, proper names and determiners to refer to entities. (ii) The spatial language understood by a robot should be full English (or another natural

Proceedings of SpLU-RoboNLP, pages 18–28
Minneapolis, MN, June 6, 2019. ©2019 Association for Computational Linguistics

language; our framework currently works only in English, but high quality translation to other languages is now available and could be integrated). (iii) For natural interaction, the major Dialog Acts used in human conversation (question, fact statement, command, opinion, Yes and No answers, etc.) should be supported. (Stolcke, 2000) (iv) The conversation should be multi-turn and (v) Interlocutors should have the ability to refer to context.

These capabilities represent advances over work such as Connell (2018) and others (listed in Bisk, 2016) where structured languages with small vocabularies and grammars specify the acceptable syntax, spanning only a small subset of full English. The rest of the paper describes key aspects of our implementation.

3 Grounding Language References to Entities

People refer to actions and entities in many different ways, using their own words. Resolving references to individual entities is a major problem in NLU, known as coreference resolution. See Elango (2006) for a survey of the domain. In particular, resolving a reference should result in a robot knowing the current position of the referred entity, so the robot can manipulate it. While object recognition is required to manipulate an object, it is clearly not sufficient: a robotics application in a large warehouse will need to process references to many thousands of objects. In a household, a robot owner will refer to people and hundreds (or even thousands) of objects. An Entity database, an innovative feature of our framework described in section 4, can resolve referring expressions in a larger applications. The Entity Database is distinct from databases containing object models used in object recognition tasks, like the ROS Household Object database (ROS.org).

A key challenge is referring to *instances* of objects - one of several individuals of the same type. When a robot is unable to resolve a particular instance (*Pick up the guitar* in a room with 3 guitars), it will ask questions like *Which guitar? The blue guitar, the black guitar, or the silver guitar?* Clicking on an instance is one way to resolve the instance and is an example of the coordination Clark (1996) referred to. However, this clicking to disambiguate may not translate well to a real application. Our framework is able to resolve object references using language, the way a person distinguishes objects of the same type: by specifying an attribute of the object, a relation to another object, an index in a list offered by the robot, an object state, or by elimination. For example:

- *the blue guitar* (color attribute)
- *the guitar on top of the bed* (spatial relation to another object)
- *the first one/ the last one* (index in a list, resolving respectively to the blue guitar or the silver guitar in the example above)
- for objects that have states, the state of an objects can be used to specify the instance desired (*the open one* – for something like a drawer or a door).
- When there are two objects of similar type, the robot will point to one of them and ask *This one? Say Yes or No.* (discrimination by elimination)

In later versions, the robot defaults to the instance closest to it, to reduce the need to clarify the instance. If a robot makes an undesirable choice, the user can say something like *go to the other one*, and a robot would move to the next instance which is spatially closest.

The examples above deal with distinguishing individuals of the same type. There are other cases requiring resolution of linguistic references (using the user's words) to entities:

- anaphoric (pronominal) references (*pick up the green briefcase, put <u>it</u> on the scanner*; 'it' refers to the green briefcase),
- *you* (the interlocutor)
- *they, them* (intelligent agents vs objects)
- a child concept from the taxonomic parent, as in *pick up the toy* (the toy giraffe, if it's the only toy in the environment),
- *here* (reference to a place in remote control situations),
- *there* (to a previously mentioned place)
- temporal time-point resolution, like *then, next, first, last* (see Perceptual Memory)

All these resolutions are supported by the Coreference Resolution module of our framework (see diagram in 8). In general, coreference resolution differentiates between ground entities (in EDB) and non-ground entities appearing in discourse (eg *I like horses*). The coreference algorithms rely on the EBD and the ontology described next.

4 The Entity Database

A key part of the solution to general grounding and coreference resolution is the Entity Database (EDB) – a database containing information about all the entities (any physical object, location) in the environment (Tellex (2011) refers to a location map for grounding). The EBD is required for the critical co-reference resolution task described above.

For a virtual simulation, the EDB can be created with tools like 3DSMax (which can import Autocad). The EDB can be exported to Excel, edited there and re-imported into a commercial Object-Oriented database used to store the Entity Database (Versant Object Database, Actian Corp). The challenge is building a EDB for the real application.

In a real application, object recognition is necessary. Databases like ROS household_objects SQL database (ROS.org), could be used in the object recognition task, which, together with the ROS database, would contribute the information needed to generate an EDB as described below, for coreference resolution purposes.

Finally, for industrial application, CAD models used in manufacturing could be used to train ML for object recognition. Majumdar (1997) explores using CAD models for object recognition.

An innovation in our framework is an extensive ontology containing **all** English nouns, built from MRD's (machine-readable dictionary, such as Wordnet). Through the ontology, knowing the type of an object (disambiguated to its sense number in the MRD) gives one access to its ontological parent, its parts, its attributes, its purpose, etc. This is a major improvement over Connell's approach which does not use an ontology. Connell's 'teaching approach' is bound to introduce problems, as humans cannot be expected to use language formally (in Connell (2018) the users are elderly people). Take Connell's example to teach a 'supporting shelf' concept: 'Supporting shelf – the LOCATION is a supporting shelf'. Ontologically, a supporting shelf is not a location, it is **in** a location. Artifacts (such as shelf) have very different properties from fixed locations. If formal reasoning were used on learnt concepts, imprecisions in definitions could have undesirable effects on robot task success (eg, if locations named differently should be different, it could be difficult for a lower shelf to be at the location of an upper shelf).

Irrespective of how the EDB is created, every object that needs to be referenced through language must be in the EDB. Objects are given common English names (eg, chair), and alternate denominations are supported, to support human references naturally. When several objects of the same type are in the same space, an instance number is appended to the type of the object to form its name. While the instance number is currently entered manually, assigning instance numbers to objects of the same type at different coordinates could be automated. In addition to a type, objects in EDB have geometric properties and parenting (on top of) scene information. In the virtual framework, attributes are given values with a tool like 3DSMax. In the real application they would be set through object recognition, and scene information. The attributes required by the virtual framework include X, Y and Z coordinates, dimensions, colors and textures, the parent (the object on top of which another object is), and a 3D position and orientation of the bounding box of the object relative to the origin of the coordinate system.

5 Spatial Relations

For natural verbal interaction, users should be able to refer to spatial relations the same way they do in everyday life. In our virtual framework, these relations are not difficult to compute, as we have the coordinates of every object in the environment, and their bounding boxes. In the virtual environment, computing a spatial relation nvolves comparing the coordinates of the relevant planes of the bounding boxes of the participating objects. This approach could be emulated in the real application, provided object recognition works well enough. Examples of spatial relations computed in this way are: X in the center of Y, X in front of Y, X to the left/right of Y, X under Y, X on Y, X behind Y, X in front of Y. An important design consideration for spatial relations is computing relations like left of Y/right of Y, in front of Y/behind Y from the point of view of the human user of the robot – as an object Y to the left of the robot will be to right of a user/operator facing that robot (or looking through a camera that faces the robot). In our framework, these relations are computed relative to the human user's camera.

At any point, a robot knows (and can tell when asked) the distance between any two objects (so in particular, the distance from itself to another object). The angle from the center of the robot's viewing frustum and another object is also computed and available in the database of observations that ground the robot event memory.

6 Knowledge Grounding and the Robot's Perceptual Memory

To interact with robots efficiently, a user needs to be able to find out what the robot knows. This comprises the commands it understands, the tasks it can execute, its 3D environment, its perception of events in this environment, and its background knowledge. In our environment, *What do you know?* is the first step in exploring a robot's knowledge. The wording of the robot's answer is intended to stimulate further interactive exploration of the knowledge (which can be vast) - by drilling down with additional questions.

Grounding means grounding basic language phrases to perceptual or motor skills, objects and locations. The grounding of objects and locations (the 3D environment) in our framework was described above. Facts known to a robot and observations made are labelled with a source of knowledge. The **sources of knowledge** are A) perception (SEE, HEAR events) – the perceptual memory is described in detail below, B) communication with other intelligent agents (a blackboard of sorts) and C) factory (static) knowledge. Commands are mapped to motor skills (see Commands).

Any robot in our framework will answer that it knows:

(i) its environment
(ii) its commands
(iii) its job
(iv) facts it was taught
(v) events it witnessed
(vi) general concepts.

These can be expanded, eg with facts about named-entities such as supported by, say, Alexa or Google (sports teams, artists, bands, movies, etc.) and world knowledge (see Future Work in 9). Categories i) – vi) are explored below.

6.1 Environment knowledge

Implementing the EDB concept in the real application will provide 'out of the box' support for the verbal interactions described below:

Spatial relations can be used in questions *What's on the table?* , or in commands *Put the vase between the candles.*

Scene contents *What do you see?* can be useful to test the vision and object recognition capabilities of remotely-controlled robots. A robot will answer a "What do you see?" question with a description of the objects in its viewing frustum.

Our framework can simulate **Robot control in remote settings**. Mediated interface to a robot can be via virtual devices such as a tablet, or cameras (in our framework, ceiling or wall cameras for interior spaces, or exterior cameras installed, for example, on an oil rig).

Number of objects in a container or area *How many cups are in the cabinet?* As certain questions like *What's in this room?* can return lengthy answers, any robot obeys *Stop* – which interrupts execution of the last command.

Questions about the **attributes** of an object in the environment *What's the height of the fridge? What's the color of the vase?* are supported directly using the EDB.

Distances in the environment *What's the distance between X and Y?* is supported using object coordinates in EDB.

Information on an object in the environment In the virtual environment, the user clicks on an object, asks *What is this?* and the robot answers with the type of the object from EDB. In a real application, the user would click on a point in the image returned by the robot's vision sensor, and the recognized object type would be used to query the EBD, using knowledge of the robot's current position and position of each object in the scene – to identify the particular instance of that object type.

6.2 Robot Commands

A robot's command set is explored with *What are your commands?* Depending on the robot, this may return *I know some movement commands, some manipulation commands, and some communication commands.* Asking *What are your (category) commands?* produces a list of the commands in that category.

Movement commands

- *Go to* <place> (Go to the bedroom). The robot will move to <place>.
- *Go to* <object> (go to the vase) the robot will turn to face <object>.
- *Face* <object> If not already facing <object>, the robot turns to face it.
- *Turn clockwise/counterclockwise (by Y degrees)*
- *Move forward/back (by Y meters)*
- *Stop* (to reset a robot).
- *Follow me, stop following me.* This command is useful in a videogame played in 3rd person, but could be changed to *follow X* (another robot).
- *go up, go down* (a robot moving on a rail is able to translate up/down or extend the manipulator arm to grab baggage from a shelf).
- *Jump* (unlikely command in a real application!)

Manipulation commands

- *Pick up* <object> (Pick up the vase). Implemented as face, reach and grab, see below. The user can ask "What do you hold?".
- *Grab* <object> (part of pick up X)
- *Drop* <object>
- *Push in* <object> (push in the cushion). *Close the drawer* works as an alternative to push in.
- *Put* <object1> *on* <object2> (put the red box on the blue box). *Put* object1 *to the left/right of* object2 (space availability is checked).
- *Put* <object1> *between* <objects> (put the vase between the candles, put the bottle between the sinks).
- *Put* <object1> *in the center of* <object2>. Knowing all dimensions enables us to check space availability prior to execution in a simpler way than in Howard (2014).
- *Rotate* <object> *by Z degrees clockwise/counterclockwise*
- *Swap* <object1> *with* <object2>. *Put* <object1> *where* <object2> *was* - also works.
- *Align* <object1> *with* <object2> (the user needs to imagine that he/she is on a plane or ship looking FORWARD, seeing a red light on his left and a green one on his right. The left (red) and right (green) and an arrow showing the forward direction of the reference object are superimposed on the reference object, and a yellow arrow is attached to the target object. The framework asks *Where should the yellow arrow point?* and differentiates two cases: when the target object is on top of the reference object, or when the two objects side by side.

- *Open door (open cupboard door* - in the kitchen); *Close the door (or the drawer, the guitar case)*

Body-part commands

- *Reach for* <object> *(part of pick up X)*
- *Point to* <object> *(or point to room)*
- *Wave*
- *Nod*

Expressing commands in different ways can be currently done with the Command Teaching facility (below). In the future, synonymic commands will be supported with semantic frames (see Future Work in 9).

Command Execution

Validation When a robot cannot execute a command (because an object is not reachable, is not movable, it is too large/small, there's not sufficient space to place an object, or because the robot is already at the destination) it will provide a diagnostic. If a command missed an argument, the robot will query for the missing argument (*go where?*).

Help A Help function is available. For complex commands, visual guidance and interactive help are available as described above for align. In Jimmy's World (see Future Work in 9) help is available conversationally.

Execution and Grounding to Motor Skills In our implementation, a robot first navigates towards the target and then turns to face it. Collision avoidance in the virtual environment is done with Havok AI, which supports 3D path planning. Collision avoidance in real applications requires sensing obstacles and avoiding them, and our movement commands could support this if necessary.

The robot moves close to the target using forward kinematics. If required, a humanoid will bend at the hips and knees while its effector starts reaching forward. This position becomes the starting position for inverse-kinematics (IK) movement of the robot effector. A similar approach that tracks state changes of objects during manipulation and after it was described in Zielinski (2015). Our framework uses HumanIK for inverse kinematics. Optimal grabbing of objects is an important area in robotic frameworks. In our framework, collision detection with Havok Physics ensures that the robot's manipulator does not go through the object is manipulates. We've implemented finger placement algorithms that rely on automatically generated 'grabbing points' (placed on opposite faces of small objects, or towards the end of larger objects) so that grabbing objects looks natural. Grabbing points can also be edited by users. Our framework supports both one-handed and two-handed grabbing of objects.

Movement to a target point, body/head rotation, effector rotation, reach and grab are basic motor skills supported in a server-side client script engine (CSE), to ground the higher level English commands listed under COMMANDS. Translating English to atomic robot commands is demonstrated in the Jan 2013 video (A.1). This approach was described as early as 2001 in Nicolescu, is used in Kress-Gazit (2008), Matuszek (2013) and in Misra (2014) [which uses pronouns without mentioning coreference resolution].

In our framework, objects have physics implemented with the commercially available Havok Physics tool (so an object falls if a robot drops it).

Teaching New Commands Required to Streamline Tasks

A robot will offer to learn a new command if it doesn't know it. Commands are entered one by one, and at execution time, they will adapt to new target objects. The initial version of Bot Colony launched in June 2014 supported learning new commands described as a sequence of existing (native) commands, where objects are parameterized. A similar approach was described subsequently in (Gemingnani, 2015).

EXAMPLE *scan the green briefc*ase The robot replies that it doesn't know 'scan', and ask if the user wants to teach it. Commands are entered one by one: *go to the shelf, pick up the green briefcase, go to the scanner, put the briefcase on the scanner, End.*

Co-reference resolution kicks in during execution to resolve the particular shelf (e.g., upper Tokyo shelf).

6.3 Robot tasks

A robot should be able to tell a user about the higher level tasks it can accomplish. Our framework treats a task like a new command, built from individual commands. Since our application was a videogame, there was no need to ground robot tasks to skills and objects, and these cannot be demonstrated in the framework. The conversation related to tasks was prototyped for use in the videogame and looks like this:

What is your job? I can clean the house, cook, wash dishes, do the laundry, babysit,...How do you clean? I vacuum the floor, I dust the furniture, I mop, etc.

However, if the user asks the robot to mop the floor, he'll learn that this function is not currently working.

6.4 Factual knowledge

For home or companion applications, knowledge of the owner and his family would enable a robot to resolve references and understand the context. Our framework supports configuring a robot with the knowledge required to serve a particular owner and family by reading in a fact base and updating EDB (it is also possible to give facts conversationally at run time). The Question Answering (QA) component can be used by a user to explore a robot's knowledge.

EXAMPLES *Who are the members of the family?Who are Ayame's children? Who is Hideki? When does X usually come home from school? What do you know about X? (Hideki is 8 years old. Hideki is the son of Ayame. Hideki goes to school). What games does Hideki enjoy? Is Masaya married? Where does Masaya work? What does the family eat for breakfast? How do you prepare X?*

Technically, these questions are not more difficult to answer than the ones Alexa or Google Assistant answer on named-entities like cities, restaurants. Conversely, if the necessary information were available, the QA component of our framework would enable a robot to fulfill functions of smart speakers, in addition to performing its physical tasks.

6.5 Perceptual Memory and Grounding of Robot Perceptions

An innovative feature of our framework is logging a robot's salient observations − events the robot witnessed- and making these accessible through question answering (QA). This is important, for example, in a security application (*When did the XYZ truck come in? When did it leave?*).

In our framework, salient observations are
- objects of interest (people, vehicle, animals − any type declared as being of interest, or OOI) entering or exiting the robot's field of view. An OOI entering/exiting the field of view triggers logging the sighting (or the speech, if applicable) for the particular type of intelligent agent or object.
- a person performing an action
- a person speaking
- any action performed by the robot

Visual and audio observations are time stamped (YYYY-MM-DD HH:MM:SS) and have a range and angle to the target. *"M. arrived on 19 August 2021 at 01:10 AM". "How do you know?" "I've seen M. from 7.2 m at an angle of 40 degrees"*. In our framework, salient observations of a robot can be played back (since we control all the actors, they are actually re-enacted on the fly).

Assigning semantics to observed actions like in "M. hid the chip in the toilet water tank" is easily done in a simulated environment, but is more challenging in a real application (how can a robot tell that someone is 'hiding' a chip?). In a real implementation, a robot could be able to recognize people and objects, and some basic actions and states of people (moving near an object, interacting with objects, sitting, lying down, coming into view, becoming not visible). Connell is proposing solution for gesture recognition in Connell (2018), but it's not clear if these can be extended to recognizing actions.

Here are some of the most useful questions supported for exploring grounding (note temporal resolution of 'first/last' 'then', 'next', 'before that', 'after that'):

- *What did X do at HH:MM on Day/Date?* (example: What *did Ayame do at 20:15 on Thursday?) What did she do **then**? What did Masaya do **next**? What did he do **before that**? How do you know that?* (grounding) *When did you first/last see X? What happened then? What happened before/after that? What happened at HH:MM on (day of week)?* (What happened at 11:30 on 26/08/2021? − this will work even if after/before don't return more facts because Jimmy the robot didn't look back/forward far enough. *When did X arrive/enter/leave the house? Where did X go after that? What did X say at (time) on (day)? What did X say before/after that? Where was X at (time) on (day of the week)/date?*

6.6 Generic Concepts

The framework supports accessing a dictionary, useful to non-native speakers of English. Intelligent conversation going beyond a definition, about any concept, requires massive knowledge about the world. In our forthcoming Jimmy's World, Jimmy (or whatever the player names his embodied bot) is able to converse on any concept and learn from the user and the community. The objective is to understand how a concept fits into everyday life. (Joseph, 2019)

7 Framework Implementation

The architecture of our NLU pipeline is shown below. The pipeline software runs on a Linux server that communicates with client software using the Google/protobuf protocol. The client manages the 3D world and robot animation, and users can interact through speech or typing. The client implements English commands sent from the server using the ground motor abilities described above. Voice input is processed by the client which calls cloud-based speech-to-text, sending the resulting text to the server-based NLU pipeline. After the pipeline generates the response, text-to-speech server-side sends audio files to the client. Language-understanding is grounded as explained in this paper. A logging service logs all interactions, and Competency (non-IDK [I don't know] answers) - is reported as a percentage of all utterances (see below). Our virtual framework represents major types of robots operating in various real environments (see images to the right). Bot Colony prototype scenes include: a home, an airport with a baggage warehouse, an oil rig, a hotel, a village filled with robotic vendors, entertainers and waiters, a hotel with robotic personnel, a manufacturing facility, a military installation, a harbour and a mine. A variety of robots are supported: humanoid robots, fixed-base greeting robots, mobile observation robots (camera bots), a rail telescopic robot, military robots, flying robots (Hunter bot) – each with commands adapted to their tasks (see Commands section).

8 Comparison with Related Work

While small vocabularies and grammars are the norm (Bisk, 2016), our pipeline supports **full** English, including idioms and phrasal verbs, in conversation. Another major novelty in our pipeline (see diagram below) is using syntactic and semantic rules mined from dictionaries for higher precision. For example, our parsing component combines the Stanford Parser, Berkeley parser and our own Template Parser, which uses syntactic rules mined from dictionaries. This parser is used to parse robot commands with very high precision (in excess of 95% on well-formed commands). On other Dialog Acts, we achieve a precision slightly superior to the component Stanford and Berkeley parsers, as we've repairing systematic parsing errors made by these parsers.

As explained below, we are currently transitioning our disambiguation to semantic frames. Coreference resolution with EDB is designed to be interactive and seek user clarification when necessary – so precision is high for entities that are in EDB.

We are logging game sessions and we compute a Competency metric (%age of utterances that don't cause *I Don't Know* answers). As players often refer to unknown entities or facts – Competency can vary widely from session to session. However, on 400 longer sessions (above 300 dialogue turns) the average Competency observed was 69%.

Figure 1: Humanoid robot

Figure 2: Telescopic rail robot

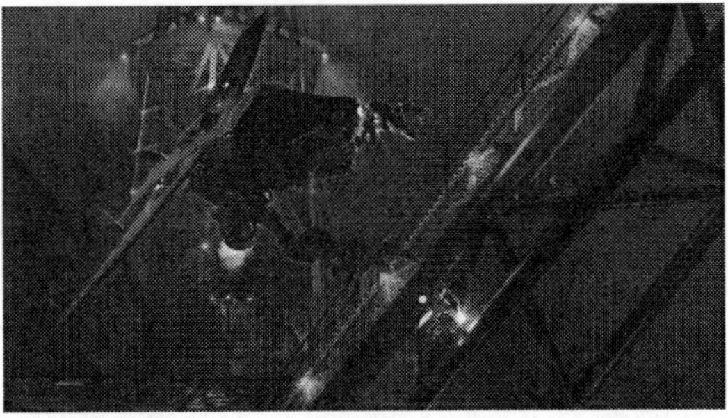

Figure 3: Airborne hunter robot

Figure 4: Underwater welding robot with welding torch and tools

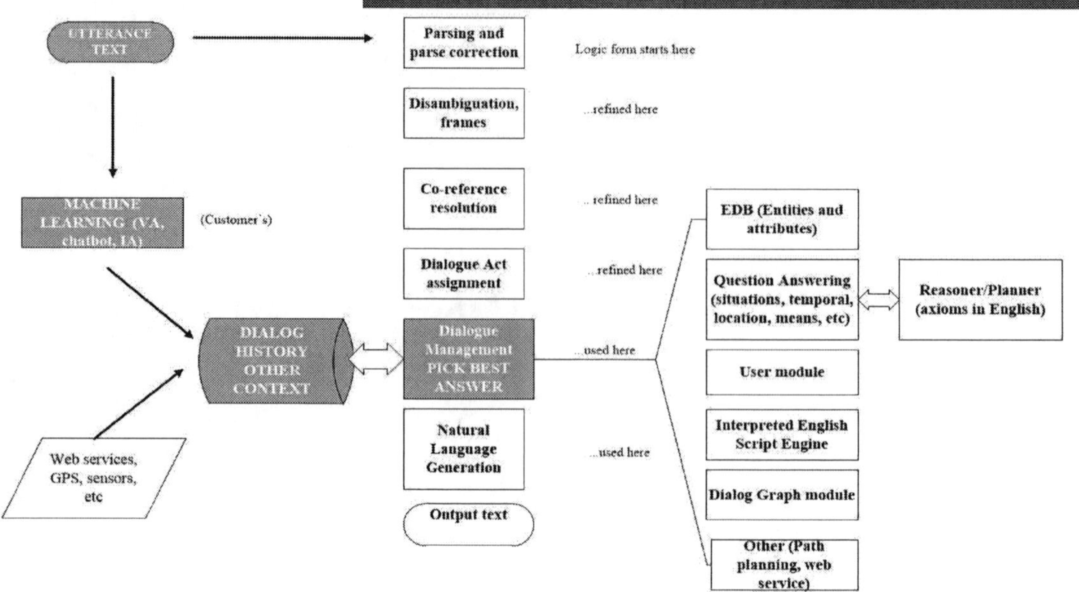

Figure 5: A key aspect of our NLU pipeline is a logic form that represents utterances formally. This logic form is initially produced from the parse tree of the input utterance. It is refined by the disambiguation module which adds sense numbers from the MRD sense inventory, and the coreference resolution module that grounds linguistic references to EDB entities. A reasoner applies axioms to this logic form, to infer, eg, that cats are born and die. Dialog Mgmt and Natural language generation also use this logic form.

9 Conclusion and Future Work

The next frontier is teaching a robot about everyday life and user preferences – a fusion between robots and intelligent assistants. This is the focus of our more recent work in Jimmy's World (Joseph, 2019). If the physical functions of a robot can be complemented with a robot ability to act as an intelligent assistant and companion - universal helper robots may become a compelling offering, especially for the elderly and people who live alone.

The Bot Colony architecture dealt with the basic issues of situated (3world based) NLU: coreference resolution, commanding robots, exploring a robot's event memory, etc. In Jimmy's World our focus is on knowledge-based NLU, so the acquisition and use of knowledge about everyday life in conversation – to cater more to personalized, intelligent assistant part of a robot's mission. A player's virtual robot in Jimmy's World will have curated knowledge from dictionaries, but will also learn from the user and the community.

Semantic frames are a way to understand language independent of the particular words and syntax used. Disambiguation to semantic frames, instead of the focus on individual Word Sense Disambiguation, is a key area of work.

A major milestone will be acquiring knowledge from individual users and the community and filtering reliable knowledge from unreliable knowledge, humour, witticism, etc.

To achieve this, we will need to refine knowledge-representation mechanisms for everyday life knowledge, and to use this knowledge in reasoning and conversation.

Since a lot of everyday life is about attaining goals and overcoming obstacles, reasoning and planning how to attain goals is another important area of work.

Machine Learning based NLU provides excellent coverage. Complementing a Machine Learning pipeline with knowledge-based NLU of the kind we are developing will result in higher precision, and deeper understanding of user utterances and is of strategic importance.

References

Yonathan Bisk, Deniz Yuret, Daniel Marcu. 2016. Natural Language Communication with Robots, NAACL.

Herbert Clark. 1995. Using Language, Google Books.

Jonathan Connell. 2018. Extensible Grounding of Speech for Robot Instruction.

J. Connell. E. Marcheret, S. Pankanti, M. Kudoh, and R. Nishiyama. 2012. *Proc. Artificial General Intelligence Conf.* (AGI-12), LNAI 7716, pp. 21-30, December 2012.

P. Elango. 2006. Coreference Resolution: A Survey, Technical Report, UW-Madison.

Gemignani, G., Bastianelli, E., Nardi, D. 2015. Teaching robots parametrized executable plans through spoken interaction. In: Proc. of AAMAS.

H. Kress-Gazit, G.E. Fainekos, and G.J. Pappas. 2008. Translating Structured English to Robot Controllers, Advanced Robotics, vol. 22, no. 12, pp. 1343–1359.

Jharna Majumdar. 1997. A CAD Model Based System for Object Recognition, Journal of Intelligent and Robotic Systems, Volume 18, Issue 4, April 1997.

Eugene Joseph. 2010. Bot Colony – A Novel Set in the Present and Near Future, North Side Inc.

Eugene Joseph. 2012. Bot Colony – a Video Game Featuring Intelligent Language-Based Interaction with the Characters, Eugene Joseph, North Side Inc., GAMNLP workshop, a special session at **JapTAL 2012** (the 8th International Conference on Natural Language Processing), Kanazawa, Japan.

Eugene Joseph. 2014. Natural Language Understanding and Text-to-Animation in Bot Colony, Gamasutra, **http://www.gamasutra.com/blogs/EugeneJoseph/20140626/219765/Natural_Language_Understanding_and_TexttoAnimation_in_Bot_Colony.php**

Eugene Joseph. 2019. Jimmy's World: Making Sense of Everyday Life References, Conversational Interaction Conference, San Jose, CA, March 11, 2019. https://docs.wixstatic.com/ugd/dbc594_ef32fd33d5b94bd9b6038f5e524c89ba.pdf

C. Matuszek, E. Herbst, L. Zettlemoyer, and D. Fox. 2013. Learning to Parse Natural Language Commands to a Robot Control System. In Experimental Robotics, pages 403–415. Springer.

M. Nicolescu and M. J. Mataric. 2001. Learning and interacting in human-robot domains. IEEE Transactions on Systems, Man, and Cybernetics, Part B,special issue on Socially Intelligent Agents - The Human in the Loop, 31(5):419–430.

Dipendra K Misra, Jaeyong Sung, Kevin Lee and Ashutosh Saxena. 2014. "Tell Me Dave: Context-Sensitive Grounding of Natural Language to Manipulation Instructions", Robotics: Science and Systems.

ROS.org The household_objects SQL database http://wiki.ros.org/household%20objects   .

Josef Ruppenhofer, Michael Ellsworth, Miriam R. L Petruck, Christopher R. Johnson, Collin F. Baker, Jan Scheffczyk. 2016. FrameNet II: Extended Theory and Practice (Revised November 1, 2016).

Andreas Stolcke, Klaus Ries et al. Dialog Act. 2000. Modeling for Automatic Tagging and Recognition of Conversational Speech Computational Linguistics, Volume 26 Number 3, ACL.

S. Tellex, T. Kollar, S. Dickerson, M.R Walter, A.G. Banerjee, S. Teller, and N. Roy. 2011. Approaching the symbol grounding problem with probabilistic graphical models. AI magazine, 32(4):64–76.

Thomas M. Howard, S. Tellex, N. Roy. 2014. A Natural Language Planner Interface for Mobile Manipulators, ICRA 2014.

Cezary Zielinski, Tomasz Komuta. 2015. An Object-Based Robot Ontology, Advances in

Intelligent Systems and Computing, January 2015.

A Appendices

A.1 Bot Colony Tech Demo Jan 2013

https://www.youtube.com/watch?v=54HpAmzaIb
s

A.2 Robot Perceptual Memory Video

https://www.youtube.com/watch?v=8zV1r8V
xWRM

Learning from Implicit Information in Natural Language Instructions for Robotic Manipulations

Ozan Arkan Can[*†]
Koç University

Pedro Zuidberg Dos Martires[*‡]
KU Leuven

Andreas Persson
Örebro University

Julian Gaal
Osnabrück University

Amy Loutfi
Örebro University

Luc De Raedt
KU Leuven

Deniz Yuret
Koç University

Alessandro Saffiotti
Örebro University

Abstract

Human-robot interaction often occurs in the form of instructions given from a human to a robot. For a robot to successfully follow instructions, a common representation of the world and objects in it should be shared between humans and the robot so that the instructions can be grounded. Achieving this representation can be done via learning, where both the world representation and the language grounding are learned simultaneously. However, in robotics this can be a difficult task due to the cost and scarcity of data. In this paper, we tackle the problem by separately learning the world representation of the robot and the language grounding. While this approach can address the challenges in getting sufficient data, it may give rise to inconsistencies between both learned components. Therefore, we further propose Bayesian learning to resolve such inconsistencies between the natural language grounding and a robot's world representation by exploiting spatio-relational information that is implicitly present in instructions given by a human. Moreover, we demonstrate the feasibility of our approach on a scenario involving a robotic arm in the physical world.

1 Introduction

Consider yourself standing in your kitchen and having your robot assist you in preparing tonight's meal. You then give it the instruction: *'fetch the bowl next to the bread knife!'*. For the robot to correctly perform your intended instruction, which is grounded in your world representation, it must correctly ground your natural language instruction into its own world representation.

This small scenario already introduces the two key components of language grounding in robotics: the construction of a world representation from sensor data and the grounding of natural language into the constructed representation. Ideally these two components would be learned in a joint fashion (Hu et al., 2017a; Johnson et al., 2017; Santoro et al., 2017; Hudson and Manning, 2018; Perez et al., 2018). However, the scarcity of data makes this approach impractical. The millions of data points necessary for state-of-the-art joint computer vision and natural language processing are simply non-existing. We opt, therefore, to separately learn the world representation component and the language grounding component.

One approach for constructing a world representation of a robot is through so-called *perceptual anchoring*. Perceptual anchoring handles the problem of creating and maintaining, over time, the correspondence between symbols in a constructed world model and perceptual data that refer to the same physical object (Coradeschi and Saffiotti, 2000). In this work, we use sensor driven bottom-up anchoring (Loutfi et al., 2005), whereby anchors (symbolic representations of objects) can be created by perceptual observations derived directly from the input sensory data. When modeling a scene, based on visual sensor data, through object anchoring, noise and uncertainties will inevitably be present. This leads, for example, to a green 'apple' object being incorrectly anchored as a 'pear'.

For the language grounding, we opt to perform the learning on **synthetic data** that simulates the world represented as anchors. This means that we do not ground the language using sensor data as

---

[*]Equal contribution
[†]ocan13@ku.edu.tr
[‡]pedro.zuidbergdosmartires@cs.kuleuven.be

Proceedings of SpLU-RoboNLP, pages 29–39
Minneapolis, MN, June 6, 2019. ©2019 Association for Computational Linguistics

signal but a symbolic representation of the world. During training these symbols are synthetic and simulated, and during the deployment of the language grounding these are anchors provided by an anchoring system. As the real world is inherently relational and as natural language instructions are often given in terms of spatial relations as well, the learned language grounder must also be able to ground spatial language such as *'next to'*.

As a result of learning the construction of a world model and the language grounding separately, **contradictions** arise **between** the world **representations** of a human and a robot. The supervision that an instruction would give to a robot is not present when learning the representation of the world of a robot. These inconsistencies then propagate through to inconsistencies between the instructions a human gives to a robot and the robot's world model. To ensure that a robot is able to correctly carry out an instruction, such inconsistencies must be resolved and the world model of the robot be matched to the world model of the human.

This is not the first paper that tackles the problem of belief revision in robotics. However, prior work (Tellex et al., 2013; Thomason et al., 2015; She and Chai, 2017), with the notable exception of (Mast et al., 2016), relied on explicit information transfer between humans and robots when inconsistencies arose in grounded language and the robot's world representation. An example would be a robot asking clarification questions until it is clear what the human meant (Tellex et al., 2013).

We propose an approach that probabilistically reasons over the grounding of an instruction and a robot's world representation in order to perform Bayesian learning to update the world representation given the grounding. This is closely related to the work of Mast et al. who also deploy a Bayesian learning approach. The key difference, however, is that they do not learn the language component but ground a description of a scene by relying on a predefined model to ground language. We demonstrate the validity of our approach for reconciling instructions and world representations on a showcase scenario involving a camera, a robot arm and a natural language interface.

2 Preliminaries

The overarching objective of our system is to plan and execute robot manipulation actions based on natural language instructions. Presumptuously,

this requires, in the first place, that both the planner of the robot manipulator, as well as the natural language grounder (cf. Section 2.2), share a joint semantically rich object-centered model of the perceived environment, i.e., a *semantic world model* (Elfring et al., 2013).

2.1 Visual Object Anchoring

In order to model a semantic object-centered representation of the external environment, we rely upon the notions and definitions found within the concept of perceptual anchoring (Coradeschi and Saffiotti, 2000). Following the approach for sensor-driven bottom-up acquisition of perceptual data, as described by (Persson et al., 2019), the used anchoring procedure is, initially, triggered by sensory input data provided by a *Kinect2 RGB-D sensor*. Each frame of input *RGB-D* data is, subsequently, processed by a *perceptual system*, which exploits both the visual 2-*D* information, as well as the 3-*D* depth information, in order to: *1)* detect and segment the subset of data (referred to as *percepts*), that originates from a single individual object in the physical world, and *2)* measure *attribute values* for each segmented percept, e.g., measuring a *position attribute* as the $\mathbb{R}^3$ geometrical center of an object, or a visual *color attribute* measured as a color histogram (in *HSV* color space).

The percept-symbol correspondence is, thereafter, established by a *symbolic system*, which handles the grounding of measured attributes values to corresponding predicate symbols through the use of *predicate grounding relations*, e.g., a certain peek in a color histogram, measured as a *color attribute*, is mapped to a corresponding predicate symbol `red`. In addition, we promote the use of an *object classification* procedure in order to semantically categorize and label each perceived object. The convolutional neural network (CNN) architecture that we use for this purpose is based on the *GoogLeNet* model (Szegedy et al., 2015), which we have trained and fine-tuned based on 101 object categories that can be expected to be found in a kitchen domain.

The extracted perceptual and symbolic information for each perceived object is then encapsulated in an internal data structure α_t^x, called an *anchor*, indexed by time t and identified by a unique identifier x (e.g. `mug-2`, `apple-4`, etc.). The goal of an *anchoring system* is to manage these anchors based on the result of a *matching function* that compares the attribute values of an unknown candidate object against the attribute values of all previously maintained anchors. Anchors are then either created or maintained through two general functionalities:

- *Acquire* – initiates a new anchor whenever a candidate object is received that does not match any existing anchor α^x.

- *Re-acquire* – extends the definition of a matching anchor α^x from time $t - k$ to time t. This functionality assures that the percepts pointed to by the anchor are the most recent perceptual (and consequently also symbolic) representation of the object.

However, comparing attribute values of anchored objects and percepts by some distance measure and deciding, based on the measure, whether an unknown object has previously been perceived or not is a non-trivial task. Nevertheless, since anchors are created or maintained through either one of the two principal functionalities *acquire* and *re-acquire*, it is evident that the desired outcome for the combined compared values is a *binary* output, i.e. should a percept be acquired or re-acquired. In previous work on anchoring (Persson et al., 2019), we have therefore suggested that the problem of invoking a correct anchoring functionality is a problem that can be approximated through learning from examples and the use of *classification algorithms*. For this work, we follow the same approach.

2.2 Natural Language Grounding

In this study, we focus on understanding spatial language that includes *pick up* and *place* related verbs, and *referring expressions*. An instruction refers to a target object using its representative features (e.g. color, shape, size). If a noun phrase does not resolve the ambiguity in the world, the instruction resolves the ambiguity by specifying the target object with its relative position to other surrounding objects. This hierarchy tries to bring the attention to finding the unique object, then shifts the attention to the targeted object. Based on this idea, we model the language grounding process as controlling the attention on the world representation by adapting the neural module networks approach proposed by Andreas et al. (2016b).

Our natural language grounder has three components: a preprocessor, an instruction parser and a program executor. Given specific anchor information (Figure 1 – № 1), the preprocessor transforms

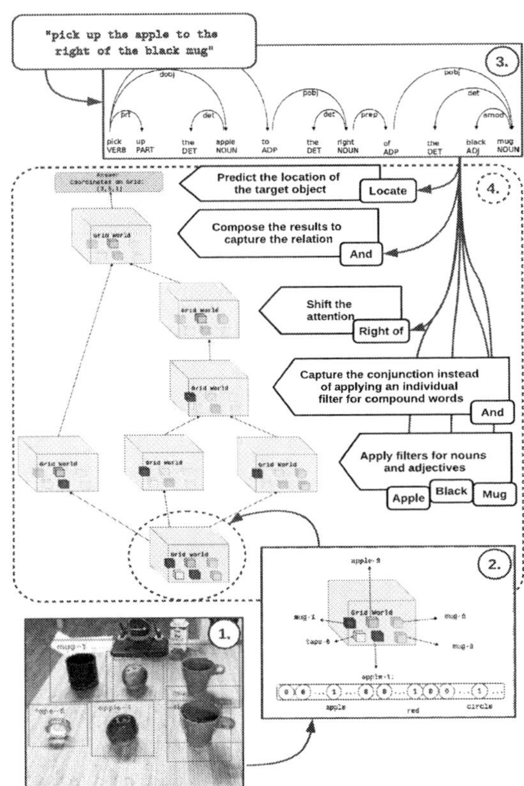

Figure 1: Demonstration of the language grounding process for the instruction "pick up the apple to the right of the black mug". Anchoring system sends the snapshot of the anchors (1). Then, a preprocessor transforms the anchors into a grid representation which the language grounding system operates on (2). The parser parses the given instruction and generates a computation graph which specifies the execution order of neural modules (3). Finally, the neural modules are executed according to the computation graph to produce the action (4).

the anchor information into an intermediate representation in grid form (Figure 1 – № 2). The instruction parser produces a computational program by exploiting the syntactic representation (Figure 1 – № 3) of the instruction with a dependency parser[1]. The program executor runs (Figure 1 – № 4) the program on the intermediate representation to produce commands.

Preprocessor. The anchoring framework maintains the object descriptions predicted from the raw visual input. To be able to ground the language onto those descriptions, we map the available information (object class, color, size and shape attributes)

---

[1] https://spacy.io/

to a 4D grid representation. We represent each anchor as a multi-hot vector and assign this vector to a cell where the real world coordinates of the object fall into.

Program Executor. The program generated by the parser is a collection of neural components that are linked to each other depending on the computation graph. The design of neural components reflects our intuition about the attention control. A ***Detect*** module is a convolutional neural network with a learnable filter that captures a noun or an adjective. This module creates an attention map over the input grid.

$$Detect(w, b, x) = relu(w \otimes x + b) \quad (1)$$

The *Detect* module operates on the original grid input x tensor, where the dimensions are (W, H, L, C). The first three dimensions represent the spatial dimensions and C denotes the length of the feature vector. w is the filter of size $(1, 1, 1, C)$ and b is the bias. $\otimes$ is a convolution operation.

Although a *Detect* module can capture the meaning of a noun phrase (e.g., red book), the model cannot generalize to unseen compound words. To overcome this, we design the *And* module to compose the output of incoming modules. This module multiplies the inputs element-wise in order to calculate the composition of words (e.g., the big red book). Since the incoming inputs are attention maps over the grid world, an ***And*** module produces a new attention map by taking the conjunction of its inputs. In the following equation, the $\odot$ denotes the element-wise multiplication.

$$And(a_1, a_2) = a_1 \odot a_2 \quad (2)$$

An output of a subgraph for a noun phrase is an attention map that highlights the positions for the corresponding objects that occur. A ***Shift*** module shifts this attention in the direction of the preposition that the module represents. This module is also a convolutional neural network similar to a ***Detect*** module. However, the module remaps the attention instead of capturing the patterns in the grid world.

$$Shift(w, a) = relu(w \otimes a) \quad (3)$$

The *Shift* module operates on an incoming attention map, where the dimensions are $(W, H, L, 1)$. w is the filter of size of $(2 * W + 1, 2 * H + 1, 2 * L + 1, 1)$. We use the padding to be able to perform the shifting operation over the whole grid. The pad size is the same as the input size.

A ***Locate*** module takes an attention map and produces a probability distribution over cells by applying a softmax classifier for being the targeted object. We use the cell with the highest probability as the prediction. A ***Position*** module gets a source anchor, a preposition and a target anchor, and produces a real world coordinate. It merely calculates the position available in the direction of the preposition from the target anchor, where the source anchor can fit.

Parser. We find the verbs in the instruction along with the subtrees attached to them. For each verb and its subtree, we search for the direct object of the verb. Then we build a subgraph for the direct object and its modifiers. Depending on the verb type, we build different subgraphs. If the verb is "pick up" related, then we look for the preposition that relates the given noun to another noun. If one is found, then a subgraph is created for the preposition object using the noun phrase that the object belongs to. Finally, the end point of the subgraph is combined with a *Shift* module. For each preposition object, we repeat the same process to handle prepositional phrase chains.

If the verb is "put" related, we find the preposition that is linked to the verb and the object of the preposition. We build a subgraph that refers to the object of the preposition similar to the "pick up" case. Finally, there is a *Position* module to produce the coordinates to put the direct object, where the position is referred with the auxiliary objects.

3   System Description

In the upper part of Figure 2, we illustrate our physical *kitchen table* system setup, which consists of the following devices: *1)* a *Kinova Jaco light-weight manipulator* (Campeau-Lecours et al., 2019), *2)* a *Microsoft Kinect2 RGB-D* sensors, and *3)* a dedicated PC with an Intel© Core™ i7-6700 processor and an Nvidia GeForce GTX 970 graphics card.

In addition, we have a modularized software architecture that utilizes the libraries and communication protocols available in the Robot Operating System (ROS)[2]. Hence, each of the modules, illustrated in the lower part of Figure 2, consists of one or several individual subsystems (or ROS nodes). For example, the *visual object anchoring* module consists of the following subsystems: 1) a *perceptual system*, 2) a *symbolic system*, and 3) an *anchoring system*. For a seamless integration be-

---

[2] http://www.ros.org/

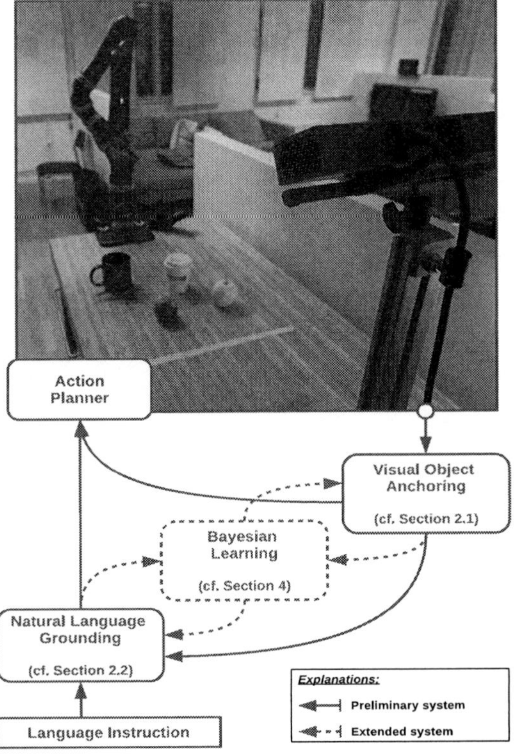

Figure 2: A depiction of both used physical system setup (upper), as well as used software architecture (lower). The arrows represent the flow of data between the modules of the software architecture. Blue solid arrows and boxes illustrate the preliminary system (outlined in Section 2), while red dashed arrows and boxes illustrate the novel extension for reasoning about different symbolic label configurations (and hence resolving inconsistencies between language and perception), by using Bayesian learning (as presented in Section 4).

tween software and hardware, we are further taking advantage of both the *MoveIt! Motion Planning Framework*[3], as well as the *ROS-Kinect2 bridge* developed by (Wiedemeyer, 2014 – 2015). The *MoveIt! "planning scene"* of the *action planner* for the robot manipulator, as well as the grid world representation used by the language grounding system (cf. Section 2.2), are, subsequently, both populated by the same updating anchoring representations (cf. Section 2.1). Hence, the visual sensory input stream is indirectly mapped to both objects considered in the dialogue by the language grounder, as well as the objects upon which actions are executed.

---

[3]https://moveit.ros.org/

4 Resolving Inconsistencies

Based purely on the perceptual input, the anchoring system produces a probability distribution $p(l)$ over the possible labels (e.g. $[0.65 : \text{apple}, 0.35 : \text{pear}]$) for each anchor. We are now interested in the probability of a label l for an anchor given a natural language instruction i and the grounding g of that instruction in the real world. This is the conditional probability $p(l \mid g, i)$. We introduce, furthermore, the notion of a label configuration c. This is easiest explained by an example: imagine having two anchors and each of the anchors has two possible labels, then there are 2×2 possible label configurations. A label configuration is, hence, a label assignment to all the anchors present in the scene.

Now we need to transform the conditional probability into a function that is computable by the anchoring system and the language grounder. The first steps (Equations 4-6) are quite straight forward and follow basic probability calculus.

$$p(l|g,i) = \sum_c p(l,c|g,i) \qquad (4)$$
$$= \sum_c p(l|c,g,i)p(c|g,i) \qquad (5)$$
$$= \sum_c p(l|c)p(c|g,i) \qquad (6)$$

In Equation 6 we assume that g and i are conditionally independent of the label of an anchor given the label configuration c. This can be seen in the following way. Imagine two anchors with two possible labels each. Given that we are in a specific label configuration, we immediately know what label the single anchors have. This means that the probability of a label for an anchor is 1 if it matches the label in the configuration and 0 otherwise. This reasoning is independent of the grounding and the instruction.

We have now split up the labels (produced by the anchoring system) and the grounding into two factors, which can be calculated separately. The first one can be calculated as follows:

$$p(l \mid c) = \frac{p(l,c)}{p(c)} = \frac{\prod_{j \in c} p(l_j)}{N_c} \qquad (7)$$

This is the product of the probabilities of the labels that constitute a label configuration divided by the number of configurations. Assuming a uniform distribution over the label configurations (division by N_c) is equivalent to assuming that each possible label configuration is equally likely *a priori*. This means that we make no assumption about which class of objects occur more regularly or which class of objects (of the 101 possible classes) occur more often together with other classes of objects.

We tackle now the second factor in Equation 6. Equation 8-11 are again straightforward probability calculus. In Equation 12 we assume that the label configuration and the instruction are independent: their probabilities factorize. In Equation 13 the probabilities of i cancel out and we assume again a uniform distribution for the label configurations (cf Equation 7). In Equation 14 we then have a numerator and denominator that are expressed in terms of $p(g \mid c, i)$, which is exactly the function approximated by our neural language grounding system, cf. subsection 2.2.

$$p(c|g,i) = \frac{p(c,g,i)}{p(g,i)} \tag{8}$$

$$= \frac{p(g|c,i)p(c,i)}{p(g,i)} \tag{9}$$

$$= \frac{p(g|c,i)p(c,i)}{\sum_c p(c,g,i)} \tag{10}$$

$$= \frac{p(g|c,i)p(c,i)}{\sum_c p(g|c,i)p(c,i)} \tag{11}$$

$$= \frac{p(g|c,i)p(c)p(i)}{\sum_c p(g|c,i)p(c)p(i)} \tag{12}$$

$$= \frac{p(g|c,i)^{1/N_c}}{\sum_c p(g|c,i)^{1/N_c}} \tag{13}$$

$$= \frac{p(g|c,i)}{\sum_c p(g|c,i)} \tag{14}$$

Plugging Equations 7 and 14 back into Equation 6 gives the learned probability of the label l of an anchor given the instruction i and the grounding g of that instruction.

$$p(l \mid g,i) = \frac{\sum_c (\prod_{j\in c} p(l_j)) p(g|c,i)}{N_c \sum_c p(g|c,i)} \tag{15}$$

As mentioned in Section 2.1 the anchoring system encapsulates 101 object categories, which means that the anchoring system produces a categorical probability distribution over 101 different labels for each anchor. With only two anchors this results in already 101^2 different configurations. It is easy to see that computing $p(l \mid g,i)$ (cf. Equation 15) suffers from this curse of dimensionality. Therefore, we limited ourselves to the two labels with the highest probability per anchor, in the experiments too. This gives 2^{N_A} possible configurations, with N_A being the number of anchors present.

5 Experiments

5.1 Synthetic Data

Data demanding nature of neural networks requires large amounts of data to generalize well. Artificial data generation is one way of generating such datasets (Andreas et al., 2016b; Kuhnle and Copestake, 2017; Johnson et al., 2016). Therefore, we designed a series of artificial learning tasks before applying the model to a real-world problem. In each task, we generate a random grid world that provides the necessary complexity and ambiguity that fit the scenario. First, an object is placed on the grid world and decorated with attributes randomly as the target object. Then depending on the scenario, an auxiliary object and distractors (objects that have similar attributes as the target object) are placed on the grid world. We also generate objects that are not related to the target (or auxiliary object) to introduce additional noise. We limit the total number of objects to 10. We set the number of distractors as 2 in the experiments. Finally, we generate the ground truth computation graph for composing neural modules. We list the scenarios below in increasing order of difficulty (i.e., a combination of the ambiguity present in the grid world and the number of language components involved).

1. Using the **name** of a targeted object in the instruction is enough to localize the targeted object.

2. There is more than one object that has the same category with a targeted object. To solve the ambiguity, one or more discriminative **adjective(s)** are used.

3. The same world configuration as the second one. To solve the ambiguity, the object is described with a **prepositional phrase** that utilizes a single referent object.

4. The same world configuration as the third. **Adjectives** are used to describe a targeted object in addition to a **prepositional phrase**. In this case, adjectives are unnecessary, but the scenario measures whether additional components bring noise or not.

5. All other objects that have the same category with a targeted object have the same set of features as the targeted object has. Hence, the targeted object is only distinguishable by its position. To solve the ambiguity, the object

is described with a **prepositional phrase** that utilizes a referent object along with necessary **adjectives**.

6. It is a **random** scenario from the above list.

5.2 Training

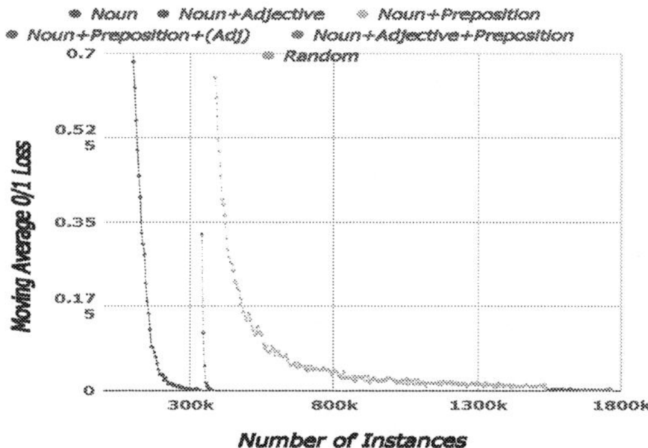

Figure 3: Learning curve of the neural modules.

To be able to measure the compositionality of learned modules, we have two different settings for the data generation. For training, we constrain the 75% of possible attributes for an object class and locations on the grid world that an instance of that object class can present. During testing, we use unconstrained samples generated for the same scenario. This way, we can evaluate if the model infers unseen word compositions, e.g. inferring red mug after seeing red book and black mug in the training time. We follow a curriculum schema to train our modules. Starting from the first scenario described in Section 5.1, we train the model on a stream of constrained randomly generated samples. We evaluate the model periodically on unconstrained samples generated for each period and continue training until the moving average error on the test data falls under a threshold (e.g. 1e-5 in our experiments). We then continue to train the model for the next scenario using learned weights. We set the number of nouns, adjectives and prepositions as 102, 26, and 27, respectively to match with the anchoring system. We use Adam (Kingma and Ba, 2014) with default parameters (i.e. $lr = 0.001$, $\beta_1 = 0.9$, $\beta_2 = 0.999$) for the optimization.

Figure 3 presents the learning curve of the model. The third graph (yellow) demonstrates that learning prepositions requires more data as compared to learning nouns (first graph) or adjectives (second graph). The reason for this behavior is twofold.

First, the *Shift* modules have more weights to be learned than the *Detect* modules. Second, while the *Detect* modules have a one to one mapping between input and output, the *Shift* modules have many to many relations. There might be more than one active area in the input of a *Shift* module. Since it needs to remap highlighted areas on the grid to other areas , it needs to see different examples that occur in different parts of the grid world in order to learn to ignore the position of the active area.

The remaining graphs show the effectiveness of our design to compose learned modules. Since we do not train any modules from scratch, we can handle the composition of nouns, adjectives and prepositions effectively. Since it is the first time we train all components together in scenario 4, the training requires more data than one would expect when compared with graphs 5 (orange) and 6 (cyan).

6 Showcase

We now proceed with a demonstration of the integrated system: we have a Kinect camera that observes the world, the anchoring representation that builds up a representation of the world based on the raw image data, the language grounder that takes as input a natural language instruction and a probabilistic reasoning component that resolves possible inconsistencies between the robot's world representation and the instruction.

The physical setup up is identical to the one depicted in the image in Figure 2: the robot arm is mounted on the opposite site of a kitchen table of the Kinect camera. The natural language instruction is passed to the language grounder via an *instruction prompt*. In each of the four panels in Figure 4, the instruction prompt is seen at the bottom as rectangular box. We further describe the scenario in the caption of Figure 4.

7 Related Work

Our work is related to two research domains: modular neural nets for language grounding and human-robot interaction for handling ambiguities in one or more modalities. Andreas et al. (2016b,a) introduced neural module networks for visual question answering. Johnson et al. (2017); Hu et al. (2017a) developed policy gradient based approaches to learn to generate layouts instead of using a dependency parser based method. Hu et al. (2017b); Yu et al. (2018); Cirik et al. (2018) applied modular neural networks approach on 'Referring Expres-

Figure 4: We give the robot the instruction: *"pick up the ball in front of the can"*. The robot executes the action and waits for further instructions. We then give the instruction to *"drop it in front of the mug"*. The problem in this step is that there is no object classified as 'mug', which means that none of the objects has as label with highest probability 'mug'. We correct for this through probabilistic reasoning over not only the top label for each object but a number of top ranked labels per object. This allows the anchoring system to correct its classification of an object based on what we as humans think an object is. Given the instruction, the anchoring system re-classifies the black object from 'pot' to 'mug'. The instruction is then successfully carried out. The recorded video can be found here: https://vimeo.com/302072685.

sion Understanding' task. Das et al. (2018) demonstrated the usage of neural module networks in decision taking in a simulated environment. To our knowledge, the present study is the first work that uses neural module networks approach in the real-world robotic setting.

Learning from human interaction has been extensively studied. Lemaignan et al. (2011, 2012) developed a cognitive architecture that makes decisions by using symbolic information provided as facts (pre-defined) or extended via human-robot dialogues. When compared with our system, their system neither operates on the sensory input nor deals with the uncertainty in the world. Tellex et al.

(2013) proposed a system to ask questions to disambiguate the ambiguities presented in the instructions. The robot decides the most ambiguous part of the command which is defined based on a metric derived from entropy and asks questions about it to reduce the uncertainty. They update the generalized grounding graph (Kollar et al., 2013) with answers obtained from the user and use these to perform inference. In contrast, we fix the ambiguity present in the perceptual data. She and Chai (2017) proposed a system to learn to ask questions during the learning of verb semantics. They work on the *Tell me Dave environment* (Misra et al., 2014). The work represents the environment as grounded state fluents (i.e. a weighted logic representation). In this work, language grounding is modeled as the difference between before and after state for an action sequence. They modeled the interactive learning as an MDP and solved it with reinforcement learning.

Thomason et al. (2015) proposed a system that learns the meaning of natural language commands through human-robot dialog. They represent the meaning of instructions with λ-calculus semantic representation. Their semantic parser starts with an initial knowledge and learns through training examples generated by the human-robot conversations. Their dialog manager is a static policy which generates questions from a discrete set of action, patient, recipient tuples. Padmakumar et al. (2017) improved this work with a learnable dialog manager. They train both the dialog manager and the semantic parser with reinforcement learning. This approach was further extended in (Thomason et al., 2019), where the authors combine the approach in Thomason et al. (2015) and Thomason et al. (2017) to obtain a system that is capable of concept acquisition through clarification dialogues. Instead of asking questions, we implicitly fix the perception with the information hidden in instructions. A further difference to these works is that we learn the language component in a simulated offline step, whereas they deploy active online learning, starting from a limited initial vocabulary.

This is also related to the work of Perera and Allen, who present a system that tries to emulate child language learning strategies by describing scenes to a robot agent, which has to learn actively new concepts. The authors deploy probabilistic reasoning to manage erroneous sensor readings in the vision system. Apart from the active learning approach, there is also a conceptual difference: in our work, we do not consider discrepencies between the perceptual system (anchoring) and the language grounder as errors in the perceptual system but simply as different models of the world.[4]

As mentioned in Section 1, the work related closest to our approach is presented in Mast et al. (2016). The authors base their work on geometric *conceptual spaces* (Gärdenfors, 2004), which situates their work in the sub-domain of *top-down anchoring* (Coradeschi and Saffiotti, 2000). The geometric conceptual spaces induce a probabilistic model-based language grounder. This enables a robot to reason probabilistically over a description of a scene, given by an other agent, and single out the object that is most likely being referred to. In contrast, we present an approach to perform Bayesian learning over a learned language grounding model and a bottom-up anchoring approach.

8 Conclusions and Future Work

We introduced the problem of belief revision in robotics based solely on implicit information available in natural language in the setting of sensor-driven bottom-up anchoring in combination with a learned language grounding model. This is in contrast to prior works, which study either explicit information or are based on top-down anchoring. We proposed a Bayesian learning approach to solve the problem and demonstrated its validity on a real world showcase involving computer vision, natural language grounding and robotic manipulation.

In future work we would like to perform a more quantitative analysis of our approach to which end it is imperative to circumvent the curse of dimensionality emerging in the Bayesian learning step (cf. Equation 15). It would also be interesting to investigate whether our approach is amenable to natural language other than instructions.

A main limitation of our current approach is the limited size of the predefined vocabulary. It would be more practicable if a robot were able to extend its vocabulary through the interaction with a human, i.e. through dialogue. A possible solution would be to learn a probabilistic model (which resolves inconsistencies between language and vision) that takes into account the possible of currently unknown vocabulary occurring. Such an approach would still allow us to learn the anchoring of objects and the language grounding separately, while learning a much richer model to resolve inconsistencies than the one described in this work.

Acknowledgements

This work has been supported by the ReGROUND project (http://reground.cs.kuleuven.be), which is a CHIST-ERA project funded by the EU H2020 framework program, the Research Foundation - Flanders, the Swedish Research Council (Vetenskapsrådet), and the Scientific and Technological Research Council of Turkey (TUBITAK). The work is also supported by Vetenskapsrådet under the grant number: 2016-05321 and by TUBITAK under the grants 114E628 and 215E201.

---

[4]This view taps into the philosophical question of whether one can ever truly know the nature of an object, cf. *thing-in-itself* (Kant, 1878), for which we omit a discussion.

References

Jacob Andreas, Marcus Rohrbach, Trevor Darrell, and Dan Klein. 2016a. Learning to compose neural networks for question answering. *arXiv preprint arXiv:1601.01705*.

Jacob Andreas, Marcus Rohrbach, Trevor Darrell, and Dan Klein. 2016b. Neural module networks. In *Proceedings of the IEEE Conference on Computer Vision and Pattern Recognition*, pages 39–48.

Alexandre Campeau-Lecours, Hugo Lamontagne, Simon Latour, Philippe Fauteux, Véronique Maheu, François Boucher, Charles Deguire, and Louis-Joseph Caron L'Ecuyer. 2019. Kinova modular robot arms for service robotics applications. In *Rapid Automation: Concepts, Methodologies, Tools, and Applications*, pages 693–719. IGI Global.

Volkan Cirik, Taylor Berg-Kirkpatrick, and Louis-Philippe Morency. 2018. Using syntax to ground referring expressions in natural images. In *Thirty-Second AAAI Conference on Artificial Intelligence*.

S. Coradeschi and A. Saffiotti. 2000. Anchoring symbols to sensor data: preliminary report. In *Proc. of the 17th AAAI Conf.*, pages 129–135, Menlo Park, CA. AAAI Press. Online at http://www.aass.oru.se/~asaffio/.

Abhishek Das, Georgia Gkioxari, Stefan Lee, Devi Parikh, and Dhruv Batra. 2018. Neural modular control for embodied question answering. *arXiv preprint arXiv:1810.11181*.

J. Elfring, S. van den Dries, M.J.G. van de Molengraft, and M. Steinbuch. 2013. Semantic world modeling using probabilistic multiple hypothesis anchoring. *Robotics and Autonomous Systems*, 61(2):95–105.

Peter Gärdenfors. 2004. *Conceptual spaces: The geometry of thought*. MIT press.

Ronghang Hu, Jacob Andreas, Marcus Rohrbach, Trevor Darrell, and Kate Saenko. 2017a. Learning to reason: End-to-end module networks for visual question answering. In *Proceedings of the IEEE International Conference on Computer Vision*, pages 804–813.

Ronghang Hu, Marcus Rohrbach, Jacob Andreas, Trevor Darrell, and Kate Saenko. 2017b. Modeling relationships in referential expressions with compositional modular networks. In *Proceedings of the IEEE Conference on Computer Vision and Pattern Recognition*, pages 1115–1124.

Drew A Hudson and Christopher D Manning. 2018. Compositional attention networks for machine reasoning. *arXiv preprint arXiv:1803.03067*.

Justin Johnson, Bharath Hariharan, Laurens van der Maaten, Li Fei-Fei, C Lawrence Zitnick, and Ross Girshick. 2016. Clevr: A diagnostic dataset for compositional language and elementary visual reasoning. *arXiv preprint arXiv:1612.06890*.

Justin Johnson, Bharath Hariharan, Laurens van der Maaten, Judy Hoffman, Li Fei-Fei, C Lawrence Zitnick, and Ross Girshick. 2017. Inferring and executing programs for visual reasoning. In *Proceedings of the IEEE International Conference on Computer Vision*, pages 2989–2998.

Immanuel Kant. 1878. *Prolegomena zu einer jeden Künftigen Metaphysik: die als Wissenschaft wird auftreten konnen*. Verlag von Leopold Voss.

Diederik Kingma and Jimmy Ba. 2014. Adam: A method for stochastic optimization. *arXiv preprint arXiv:1412.6980*.

Thomas Kollar, Stefanie Tellex, Matthew R Walter, Albert Huang, Abraham Bachrach, Sachi Hemachandra, Emma Brunskill, Ashis Banerjee, Deb Roy, Seth Teller, et al. 2013. Generalized grounding graphs: A probabilistic framework for understanding grounded language. *JAIR*.

Alexander Kuhnle and Ann Copestake. 2017. Shapeworld-a new test methodology for multimodal language understanding. *arXiv preprint arXiv:1704.04517*.

Séverin Lemaignan, Raquel Ros, Rachid Alami, and Michael Beetz. 2011. What are you talking about? grounding dialogue in a perspective-aware robotic architecture. In *2011 RO-MAN*, pages 107–112. IEEE.

Séverin Lemaignan, Raquel Ros, E Akin Sisbot, Rachid Alami, and Michael Beetz. 2012. Grounding the interaction: Anchoring situated discourse in everyday human-robot interaction. *International Journal of Social Robotics*, 4(2):181–199.

A. Loutfi, S. Coradeschi, and A. Saffiotti. 2005. Maintaining coherent perceptual information using anchoring. In *Proc. of the 19th IJCAI Conf.*, pages 1477–1482, Edinburgh, UK.

Vivien Mast, Zoe Falomir, and Diedrich Wolter. 2016. Probabilistic reference and grounding with pragr for dialogues with robots. *Journal of Experimental & Theoretical Artificial Intelligence*, 28(5):889–911.

Dipendra K Misra, Jaeyong Sung, Kevin Lee, and Ashutosh Saxena. 2014. Tell me dave: Contextsensitive grounding of natural language to mobile manipulation instructions. In *in RSS*. Citeseer.

Aishwarya Padmakumar, Jesse Thomason, and Raymond J Mooney. 2017. Integrated learning of dialog strategies and semantic parsing. In *Proceedings of the 15th Conference of the European Chapter of the Association for Computational Linguistics (EACL)*.

Ian Perera and James Allen. 2015. Quantity, contrast, and convention in cross-situated language comprehension. In *Proceedings of the Nineteenth Conference on Computational Natural Language Learning*, pages 226–236.

Ethan Perez, Florian Strub, Harm De Vries, Vincent Dumoulin, and Aaron Courville. 2018. Film: Visual reasoning with a general conditioning layer. In *Thirty-Second AAAI Conference on Artificial Intelligence*.

Andreas Persson, Pedro Zuidberg Dos Martires, Amy Loutfi, and Luc De Raedt. 2019. Semantic relational object tracking. *arXiv preprint arXiv:1902.09937.*

Adam Santoro, David Raposo, David G Barrett, Mateusz Malinowski, Razvan Pascanu, Peter Battaglia, and Timothy Lillicrap. 2017. A simple neural network module for relational reasoning. In *Advances in neural information processing systems*, pages 4967–4976.

Lanbo She and Joyce Chai. 2017. Interactive learning of grounded verb semantics towards human-robot communication. In *Proceedings of the 55th Annual Meeting of the Association for Computational Linguistics (Volume 1: Long Papers)*, volume 1, pages 1634–1644.

Christian Szegedy, Wei Liu, Yangqing Jia, Pierre Sermanet, Scott Reed, Dragomir Anguelov, Dumitru Erhan, Vincent Vanhoucke, and Andrew Rabinovich. 2015. Going deeper with convolutions. In *Proceedings of the IEEE Conference on Computer Vision and Pattern Recognition*, pages 1–9.

Stefanie Tellex, Pratiksha Thakerll, Robin Deitsl, Dimitar Simeonovl, Thomas Kollar, and Nicholas Royl. 2013. Toward information theoretic human-robot dialog. *Robotics*, page 409.

Jesse Thomason, Aishwarya Padmakumar, Jivko Sinapov, Justin Hart, Peter Stone, and Raymond J Mooney. 2017. Opportunistic active learning for grounding natural language descriptions. In *Conference on Robot Learning*, pages 67–76.

Jesse Thomason, Aishwarya Padmakumar, Jivko Sinapov, Nick Walker, Yuqian Jiang, Harel Yedidsion, Justin Hart, Peter Stone, and Raymond J Mooney. 2019. Improving grounded natural language understanding through human-robot dialog. *arXiv preprint arXiv:1903.00122.*

Jesse Thomason, Shiqi Zhang, Raymond J Mooney, and Peter Stone. 2015. Learning to interpret natural language commands through human-robot dialog. In *IJCAI*, pages 1923–1929.

Thiemo Wiedemeyer. 2014 – 2015. IAI Kinect2. `https://github.com/code-iai/iai_kinect2`. Accessed February 28, 2019.

Licheng Yu, Zhe Lin, Xiaohui Shen, Jimei Yang, Xin Lu, Mohit Bansal, and Tamara L Berg. 2018. Mattnet: Modular attention network for referring expression comprehension. In *Proceedings of the IEEE Conference on Computer Vision and Pattern Recognition*, pages 1307–1315.

Multi-modal Discriminative Model for Vision-and-Language Navigation

Haoshuo Huang[*]   **Vihan Jain**[*]   **Harsh Mehta**   **Jason Baldridge**   **Eugene Ie**
Google AI Language
{haoshuo, vihan, harshm, jridge, eugeneie}@google.com

Abstract

Vision-and-Language Navigation (VLN) is a natural language grounding task where agents have to interpret natural language instructions in the context of visual scenes in a dynamic environment to achieve prescribed navigation goals. Successful agents must have the ability to parse natural language of varying linguistic styles, ground them in potentially unfamiliar scenes, plan and react with ambiguous environmental feedback. Generalization ability is limited by the amount of human annotated data. In particular, *paired* vision-language sequence data is expensive to collect. We develop a discriminator that evaluates how well an instruction explains a given path in VLN task using multi-modal alignment. Our study reveals that only a small fraction of the high-quality augmented data from Fried et al. (2018), as scored by our discriminator, is useful for training VLN agents with similar performance on previously unseen environments. We also show that a VLN agent warm-started with pre-trained components from the discriminator outperforms the benchmark success rates of 35.5 by 10% relative measure on previously unseen environments.

1   Introduction

There is an increased research interest in the problems containing multiple modalities (Yu and Siskind, 2013; Chen et al., 2015; Vinyals et al., 2017; Harwath et al., 2018). The models trained on such problems learn similar representations for related concepts in different modalities. Model components can be pretrained on datasets with individual modalities, the final system must be trained (or fine-tuned) on task-specific datasets (Girshick et al., 2014; Zeiler and Fergus, 2014).

In this paper, we focus on vision-and-language navigation (VLN), which involves understanding visual-spatial relations as described in instructions written in natural language. In the past, VLN datasets were built on virtual environments, with MacMahon et al. (2006) being perhaps the most prominent example. More recently, challenging photo-realistic datasets containing instructions for paths in real-world environments have been released (Anderson et al., 2018b; de Vries et al., 2018; Chen et al., 2018). Such datasets require annotations by people who follow and describe paths in the environment. Because the task is quite involved–especially when the paths are longer–obtaining human labeled examples at scale is challenging. For instance, the Touchdown dataset (Chen et al., 2018) has only 9,326 examples of the complete task. Others, such as Cirik et al. (2018) and Hermann et al. (2019) side-step this problem by using formulaic instructions provided by mapping applications. This makes it easy to get instructions at scale. However, since these are not *natural* language instructions, they lack the quasi-regularity, diversity, richness and errors inherent in how people give directions. More importantly, they lack the more interesting connections between language and the visual scenes encountered on a path, such as *head over the train tracks, hang a right just past a cluster of palm trees and stop by the red brick town home with a flag over its door.*

In general, the performance of trained neural models is highly dependent on the amount of available training data. Since human-annotated data is expensive to collect, it is imperative to maximally exploit existing resources to train models that can be used to improve the navigation agents. For instance, to extend the Room-to-Room (R2R) dataset (Anderson et al., 2018b), Fried et al. (2018) created an augmented set of instructions for randomly generated paths in the same underlying environment. These instructions were generated by a speaker model that was trained on the available

---

*Authors contributed equally.

Proceedings of SpLU-RoboNLP, pages 40–49
Minneapolis, MN, June 6, 2019. ©2019 Association for Computational Linguistics

human-annotated instructions in R2R. Using this augmented data improved the navigation models in the original paper as well as later models such as Wang et al. (2018a). However, our own inspection of the generated instructions revealed that many have little connection between the instructions and the path they were meant to describe, raising questions about what models can and should learn from noisy, automatically generated instructions.

We instead pursue another, high precision strategy for augmenting the data. Having access to an environment provides opportunities for creating instruction-path pairs for modeling alignments. In particular, given a path and a navigation instruction created by a person, it is easy to create incorrect paths by creating permutations of the original path. For example, we can hold the instructions fixed, but reverse or shuffle the sequence of perceptual inputs, or sample random paths, including those that share the start or end points of the original one. Crucially, given the diversity and relative uniqueness of the properties of different rooms and the trajectories of different paths, it is highly unlikely that the original instruction will correspond well to the mined negative paths.

This negative path mining strategy stands in stark contrast with approaches that create new instructions. Though they cannot be used to directly train navigation agents, negative paths can instead be used to train discriminative models that can assess the fit of an instruction and a path. As such, they can be used to judge the quality of machine-generated extensions to VLN datasets and possibly reject bad instruction-path pairs. More importantly, the components of discriminative models can be used for initializing navigation models themselves and thus allow them to make more effective use of the limited positive paths available.

We present four main contributions. First, we propose a *discriminator* model (Figure 1) that can predict how well a given instruction explains the paired path. We list several cheap negative sampling techniques to make the discriminator more robust. Second, we show that only a small portion of the augmented data in Fried et al. (2018) are high fidelity. Including just a small fraction of them in training is sufficient for reaping most of the gains afforded by the full augmentation set: using just the top 1% augmented data samples, as scored by the discriminator, is sufficient to generalize to previously unseen environments. Third, we train

the discriminator using *alignment-based* similarity metric that enables the model to align same concepts in the language and visual modalities. We provide a qualitative assessment of the alignment learned by the model. Finally, we show that a navigation agent, when initialized with components of fully-trained discriminator, outperforms the existing benchmark on success rate by over 10% relative measure on previously unseen environments.

2   The Room-to-Room Dataset

Room-to-Room (R2R) is a visually-grounded natural language navigation dataset in photo-realistic environments (Anderson et al., 2018b). Each environment is defined by a graph where nodes are locations with egocentric panoramic images and edges define valid connections for agent navigation. The navigation dataset consists of language instructions paired with reference paths, where each path is defined by a sequence of graph nodes. The data collection process is based on sampling pairs of start/end nodes and defining the shortest path between the two. Furthermore the collection process ensures no paths are shorter than 5m and must be between 4 to 6 edges. Each sampled path is associated with 3 natural language instructions collected from Amazon Mechanical Turk with an average length of 29 tokens from a vocabulary of 3.1k tokens. Apart from the training set, the dataset includes two validation sets and a test set. One of the validation sets includes new instructions on environments overlapping with the training set (Validation Seen), and the other is entirely disjoint from the training set (Validation Unseen).

Several metrics are commonly used to evaluate agents' ability to follow navigation instructions. *Path Length (PL)* measures the total length of the predicted path, where the optimal value is the length of the reference path. *Navigation Error (NE)* measures the distance between the last nodes in the predicted path and the reference path. *Success Rate (SR)* measures how often the last node in the predicted path is within some threshold distance d_{th} of the last node in the reference path. More recently, Anderson et al. (2018a) proposed the *Success weighted by Path Length (SPL)* measure that also considers whether the success criteria was met (i.e., whether the last node in the predicted path is within some threshold d_{th} of the reference path) and the normalized path length. Agents should minimize NE and maximize SR and SPL.

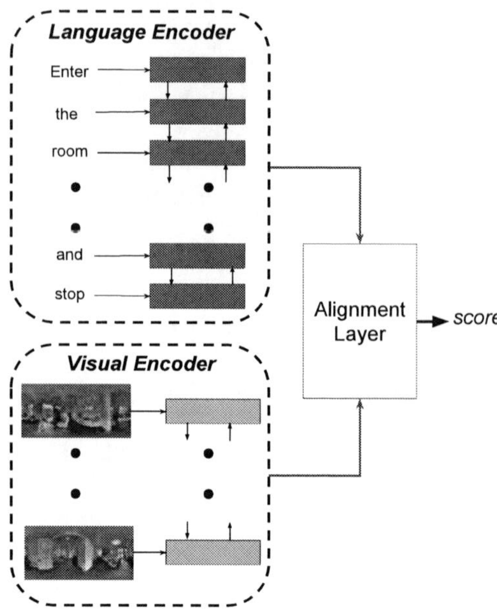

Figure 1: Overview of the discriminator model structure. Alignment layer corresponds to Eq.5,6,7

3 Discriminator Model

VLN tasks are composed of instruction-path pairs, where a path is a sequence of connected locations along with their corresponding perceptual contexts in some environment. While the core task is to create agents that can follow the navigation instructions to reproduce estimates of reference paths, we instead explore models that focus on the simpler problem of judging whether an instruction-path pair are a good match for one another. These models would be useful in measuring the quality of machine-generated instruction-path pairs. Another reasonable expectation from such models would be that they are also able to align similar concepts in the two modalities, e.g., in an instruction like *"Turn right and move forward around the bed, enter the bathroom and wait there."*, it is expected that the word *bed* is better aligned with a location on the path that has a bed in the agent's egocentric view.

To this effect, we train a discriminator model that learns to delineate positive instruction-path pairs from negative pairs sampled using different strategies described in Sec.3.2. The discrimination is based on an alignment-based similarity score that determines how well the two input sequences align. This encourages the model to map perceptual and textual signals for final discrimination.

3.1 Model Structure

We use a two-tower architecture to independently encode the two sequences, with one tower encoding the token sequence $x_1, x_2, ..., x_n$ in the instruction $\mathcal{X}$ and another tower encoding the visual input sequence $v_1, v_2, ..., v_m$ from the path $\mathcal{V}$. Each tower is a bi-directional LSTM (Schuster and Paliwal, 1997) which constructs the latent space representation H of a sequence $i_1, i_2, ..., i_k$ following:

$$H = [h_1; h_2; ...; h_k] \tag{1}$$

$$h_t = g(\overrightarrow{h}_t, \overleftarrow{h}_t) \tag{2}$$

$$\overrightarrow{h}_t = LSTM(i_t, \overrightarrow{h}_{t-1}) \tag{3}$$

$$\overleftarrow{h}_t = LSTM(i_t, \overleftarrow{h}_{t+1}) \tag{4}$$

where g function is used to combine the output of forward and backward LSTM layers. In our implementation, g is the concatenation operator.

We denote the latent space representation of instruction $\mathcal{X}$ as H^X and path $\mathcal{V}$ as H^V and compute the alignment-based similarity score as following:

$$A = H^X (H^V)^T \tag{5}$$

$$\{c\}_{l=1}^{l=X} = \text{softmax}(A^l) \cdot A^l \tag{6}$$

$$\text{score} = \text{softmin}(\{c\}_{l=1}^{l=X}) \cdot \{c\}_{l=1}^{l=X} \tag{7}$$

where $(.)^T$ is matrix transpose transformation, A is the alignment matrix whose dimensions are $[n, m]$, A^l is the l-th row vector in A and softmin(Z) = $\frac{\exp^{-Z_j}}{\sum \exp^{-Z_j}}$. Eq.6 corresponds to taking a softmax along the columns and summing the columns, which amounts to content-based pooling across columns. Then we apply softmin operation along the rows and sum the rows up to get a scalar in Eq.7. Intuitively, optimizing this score encourages the learning algorithm to construct the best worst-case sequence alignment between the two input sequences in latent space.

3.2 Training

Training data consists of instruction-path pairs which may be similar (positives) or dissimilar (negatives). The training objective maximizes the log-likelihood of predicting higher alignment-based similarity scores for similar pairs.

We use the human annotated demonstrations in the R2R dataset as our positives and explore three strategies for sampling negatives. For a given

Learning	PS	PR	RW	AUC
no-curriculum	✓			64.5
no-curriculum		✓		60.5
no-curriculum			✓	63.1
no-curriculum	✓	✓		72.1
no-curriculum		✓	✓	66.0
no-curriculum	✓		✓	70.8
no-curriculum	✓	✓	✓	72.0
curriculum	✓	✓	✓	**76.2**

Table 1: Results on training in different combinations of datasets and evaluating against validation dataset containing PR and RW negatives only.

instruction-path pair, we sample negatives by keeping the same instruction but altering the path sequence by:

- *Path Substitution (PS)* – randomly picking other paths from the same environment as negatives.

- *Partial Reordering (PR)* – keeping the first and last nodes in the path unaltered and shuffling the intermediate locations of the path.

- *Random Walks (RW)* – sampling random paths of the same length as the original path that either (1) start at the same location and end sufficiently far from the original path or (2) end at the same location and start sufficiently far from the original path.

4 Results

Our experiments are conducted using the R2R dataset (Anderson et al., 2018b). Recently, Fried et al. (2018) introduced an augmented dataset (referred to as `Fried-Augmented` from now on) that is generated by using a speaker model and they show that the models trained with both the original data and the machine-generated augmented data improves agent success rates.

We show three main results. First, the discriminator effectively differentiates between high-quality and low-quality paths in `Fried-Augmented`. Second, we rank all instruction-path pairs in `Fried-Augmented` with the discriminator and train with a small fraction judged to be the highest quality—using just the top 1% to 5% (the highest quality pairs) provides most of the benefits derived from the entirety of `Fried-Augmented` when

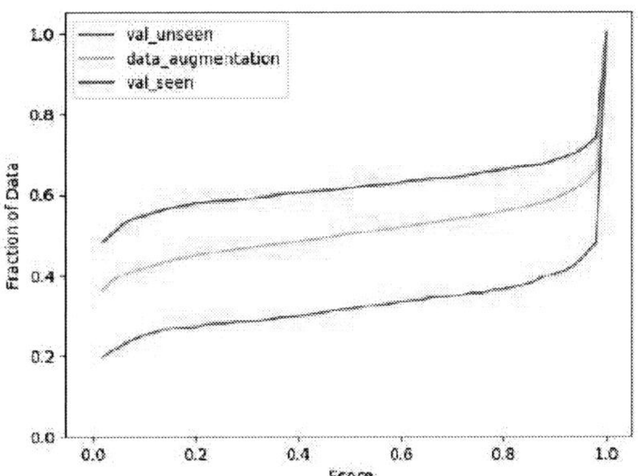

Figure 2: Culmulative distributions of discriminator scores for different datasets. The mean of distribution for R2R *validation seen*, `Fried-Augmented` and R2R *validation unseen* is 0.679, 0.501, and 0.382 respectively.

generalizing to previously unseen environments. Finally, we initialize a navigation agent with the visual and language components of the trained discriminator. This strategy allows the agent to benefit from the discriminator's multi-modal alignment capability and more effectively learn from the human-annotated instructions. This agent outperforms existing benchmarks on previously unseen environments as a result.

4.1 Discriminator Results

We create two kinds of dataset for each of the negative sampling strategies defined in Section 3.2 – a training set from paths in R2R *train* split and validation set from paths in R2R *validation seen* and *validation unseen* splits. The area-under ROC curve (AUC) is used as the evaluation metric for the discriminator. From preliminary studies, we observed that the discriminator trained on dataset containing PS negatives achieved AUC of 83% on validation a dataset containing PS negatives only, but fails to generalize to validation set containing PR and PW negatives (AUC of 64.5%). This is because it is easy to score PS negatives by just attending to first or last locations, while scoring PR and PW negatives may require carefully aligning the full sequence pair. Therefore, to keep the task challenging, the validation set was limited to contain validation splits from PR and RW negative sampling strategies only. Table 1 shows the results of training the discriminator using various combinations of negative sampling.

Generally, training the discriminator with PS

Dataset	Score	Example
Fried-Augmented	0.001	*Gym* *Walk out of the bedroom and turn left. Walk past the couch and turn right. Walk past the pool and stop on the second step.*
	0.999	*Go up the stairs and turn left. Walk past the kitchen and dining table. Stop behind the dining table.*
Validation Seen	0.014	*Hardly visible* *Walk across the patio, stop at hanging basket chair.*
	0.999	*Leave the closet and take a right into the hallway. In the hall walk straight and stick left passing a cabinet on your left. Once past the cabinet go into the first room on your left.*
Validation Unseen	0.00004	*Exit the room then turn left and go up the steps then turn right and turn right and wait near the beige couches.*
	0.9808	*Walk down the stairs, at the landing enter the second doorway on the left. Wait near the bookshelf.*

Table 2: Selected samples from datasets with discriminator scores.

negatives helps model performance across the board. Simple mismatch patterns in PS negatives help bootstrap the model with a good initial policy for further fine-tuning on tougher negatives patterns in PR and RW variations. For example in PS negatives, a path that starts in a bathroom does not match with an instruction that begins with "*Exit the bedroom.*"–this would be an easy discrimination pair. In contrast, learning from just PR and RW negatives fails to reach similar performance. To further confirm this hypothesis, we train a discriminator using curriculum learning (Bengio et al., 2009) where the model is first trained on only PS negatives and then fine-tuned on PR and RW negatives. This strategy outperforms all others, and the resulting best performing discriminator is used for conducting studies in the following subsections.

Discriminator Score Distribution Fig.2 shows the discriminator's score distribution on different R2R datasets. Since Fried-Augmented contains paths from houses seen during training, it would be expected that discriminator's scores on *validation seen* and Fried-Augmented datasets be the same if the data quality is comparable. However there is a clear gap in the discriminator's con-

fidence between the two datasets. This matches our subjective analysis of Fried-Augmented where we observed many paths had clear starting/ending descriptions but the middle sections were often garbled and had little connection to the perceptual path being described. Table 2 contains some samples with corresponding discriminator scores.

Finally we note that the discriminator scores on *validation unseen* are rather conservative even though the model differentiates between positives and negatives from validation set reasonably well (last row in Table 1).

4.2 Training Navigation Agent

We conducted studies on various approaches to incorporate selected samples from Fried-Augmented to train navigation agents and measure their impact on agent navigation performance. The studies illustrate that navigation agents have higher success rates when they are trained on higher-quality data (identified by discriminator) with sufficient diversity (introduced by random sampling). When the agents are trained with mixing selected samples from

Dataset size	Strategy	PL		NE ↓		SR ↑		SPL ↑	
		U	S	U	S	U	S	U	S
1%	Top	11.2	11.1	8.5	8.5	20.4	21.2	16.6	**17.6**
	Bottom	10.8	10.7	8.9	9.0	15.4	16.3	14.1	13.1
	Random Full	11.7	12.5	8.1	8.3	22.1	21.2	**17.9**	16.6
	Random Bottom	14.2	15.8	8.4	8.1	19.7	21.7	14.3	15.6
	Random Top	15.9	15.6	**7.9**	**7.6**	**22.6**	**25.4**	15.2	14.8
2%	Top	11.3	11.7	8.2	7.9	22.3	25.5	18.5	21.0
	Bottom	11.4	14.5	8.4	9.1	17.5	17.7	14.1	12.7
	Random Full	13.3	10.8	7.9	7.9	24.3	25.5	18.2	**22.7**
	Random Bottom	15.2	18.2	8.1	8.1	20.5	20.8	11.8	16.0
	Random Top	12.9	14.0	**7.6**	**7.5**	**25.6**	**25.8**	**19.5**	19.7
5%	Top	17.6	16.9	7.7	7.2	24.6	28.2	14.4	18.2
	Bottom	10.0	10.2	8.3	8.2	20.1	23.2	**17.1**	19.4
	Random Full	17.8	21.4	7.3	7.0	27.2	29.1	16.4	14.3
	Random Bottom	16.3	10.4	7.9	8.3	22.1	23.0	14.2	20.1
	Random Top	20.0	15.0	**7.0**	**6.9**	**27.7**	**30.6**	14.8	**22.1**

Table 3: Results on R2R validation unseen paths (U) and seen paths (S) when trained only with small fraction of `Fried-Augmented` ordered by discriminator scores. For *Random Full* study, examples are sampled uniformly over entire dataset. For *Random Top/Bottom* study, examples are sampled from top/bottom 40% of ordered dataset. SPL and SR are reported as percentages and NE and PL in meters.

`Fried-Augmented` to **R2R** *train* dataset, only the top 1% from `Fried-Augmented` is needed to match the performance on existing benchmarks.

Training Setup. The training setup of the navigation agent is identical to Fried et al. (2018). The agent learns to map the natural language instruction $\mathcal{X}$ and the initial visual scene v_1 to a sequence of actions $a_{1..T}$. Language instructions $\mathcal{X} = x_{1..n}$ are initialized with pre-trained GloVe word embeddings (Pennington et al., 2014) and encoded using a bidirectional RNN (Schuster and Paliwal, 1997). At each time step t, the agent perceives a 360-degree panoramic view of its surroundings from the current location. The view is discretized into m view angles ($m = 36$ in our implementation, 3 elevations x 12 headings at 30-degree intervals). The image at view angle i, heading angle ϕ and elevation angle θ is represented by a concatenation of the pre-trained CNN image features with the 4-dimensional orientation feature [sin ϕ; cos ϕ; sin θ; cos θ] to form $v_{t,i}$. As in Fried et al. (2018), the agent is trained using *student forcing* where actions are sampled from the model during training, and supervised using a shortest-path action to reach the goal state.

Training using `Fried-Augmented` only. The experiments in Table 3 are based on training a navigation agent on different fractions of the `Fried-Augmented` dataset (X={1%, 2%, 5%}) and sampling from different parts of the discriminator score distribution (*Top, Bottom, Random Full, Random Top, Random Bottom*). The trained agents are evaluated on both *validation seen* and *validation unseen* datasets.

Not surprisingly, the agents trained on examples sampled from the *Top* score distribution consistently outperform the agents trained on examples from the *Bottom* score distribution. Interestingly, the agents trained using the *Random Full* samples is slightly better than agents trained using just the *Top* samples. This suggests that the agent benefits from higher diversity samples. This is confirmed by the study *Random Top* where the agents trained using high quality samples with sufficient diversity consistently outperform all other agents.

Training using both R2R *train* and `Fried-Augmented`. To further investigate the utility of the discriminator, the navigation agent is trained with the full R2R *train* dataset (which contains human annotated data) as well as selected fractions of `Fried-Augmented`[1]. Table 4 shows the results.

---

[1] We tried training on `Fried-Augmented` first and then fine-tuning on R2R *train* dataset, as done in Fried et al. (2018), but didn't find any appreciable difference in agent's performance in any of the experiments.

Dataset	PL		NE $\downarrow$		SR $\uparrow$		SPL $\uparrow$	
	U	S	U	S	U	S	U	S
Benchmark[2]	-	-	6.6	3.36	35.5	66.4	-	-
0%	17.8	18.5	6.8	5.3	32.1	46.1	21.9	30.3
1%	12.5	11.2	6.4	5.7	35.2	45.3	28.9	39.1
2%	14.5	15.1	6.5	5.5	35.7	44.6	27.0	34.1
5%	17.0	12.9	**6.1**	5.6	36.0	44.8	23.6	37.0
40%	14.9	11.9	6.4	5.5	**36.5**	49.1	27.1	43.4
60%	16.8	15.7	6.3	5.3	36.0	47.2	24.7	35.4
80%	17.1	18.5	6.2	5.2	35.8	45.0	23.8	29.6
100%	15.6	15.9	6.4	**4.9**	36.0	**51.9**	**29.0**	**43.0**

Table 4: Results[3] on R2R validation unseen (U) and validation seen (S) paths when trained with full training set and selected fraction of `Fried-Augmented`. SPL and SR are reported as percentages and NE and PL in meters.

Method	Split	PL	NE $\downarrow$	SR $\uparrow$	SPL $\uparrow$
Speaker-Follower model (Fried et al., 2018)	U	-	6.6	35.5	-
	S	-	**3.36**	**66.4**	-
Speaker-Follower model (our implementation)	U	15.6	6.4	36.0	29.0
	S	15.9	4.9	51.9	43.0
Our implementation, using discriminator pre-training	U	16.7	**5.9**	**39.1**	26.8
	S	15.4	5.0	50.4	39.1

Table 5: Results on R2R validation unseen (U) and validation seen (S) paths after initializing navigation agent's instruction and visual encoders with discriminator.

Validation Unseen: The performance of the agents trained with just 1% `Fried-Augmented` matches with benchmark for NE and SR. With just 5% `Fried-Augmented`, the agent starts outperforming the benchmark for NE and SR. Since `Fried-Augmented` was generated by a speaker model that was trained on R2R *train*, the language diversity in the dataset is limited, as evidenced by the unique token count: R2R *train* has 2,602 unique tokens while `Fried-Augmented` has only unique 369 tokens. The studies show that only a small fraction of top scored `Fried-Augmented` is needed to augment R2R *train* to achieve the full performance gain over the benchmark.

Validation Seen: Since `Fried-Augmented` contains paths from houses seen during training, mixing more of it with R2R *train* helps the agent overfit on *validation seen*. Indeed, the model's performance increases nearly monotonically on NE and SR as higher fraction of `Fried-Augmented` is mixed in the training data. The agent performs best when it is trained on all of `Fried-Augmented`.

Initializing with Discriminator. To further demonstrate the usefulness of the discriminator strategy, we initialize a navigation agent's instruction and visual encoder using the discriminator's instruction and visual encoder respectively. We note here that since the navigation agent encodes the visual input sequence using LSTM, we re-train the best performing discriminator model using LSTM (instead of bidirectional-LSTM) visual encoder so that the learned representations can be transferred correctly without any loss of information. We observed a minor degradation in the performance of the modified discriminator. The navigation agent so initialized is then trained as usual using *student*

---

[2]For a fair comparison, the benchmark is the Speaker-Follower model from Fried et al. (2018) which uses panoramic action space and augmented data, but no beam search (pragmatic inference).

[3]Our results of the agents trained on the full R2R *train* and 100% `Fried-Augmented` match with Speaker-Follower benchmark on validation unseen but are lower on validation seen. This is likely due to differences in model capacity, hyper-parameter choices and image features used in our implementation. The image features used in our implementation are obtained through a convolutional network trained with a semantic ranking objective on a proprietary image dataset with over 100+ million images (Wang et al., 2014).

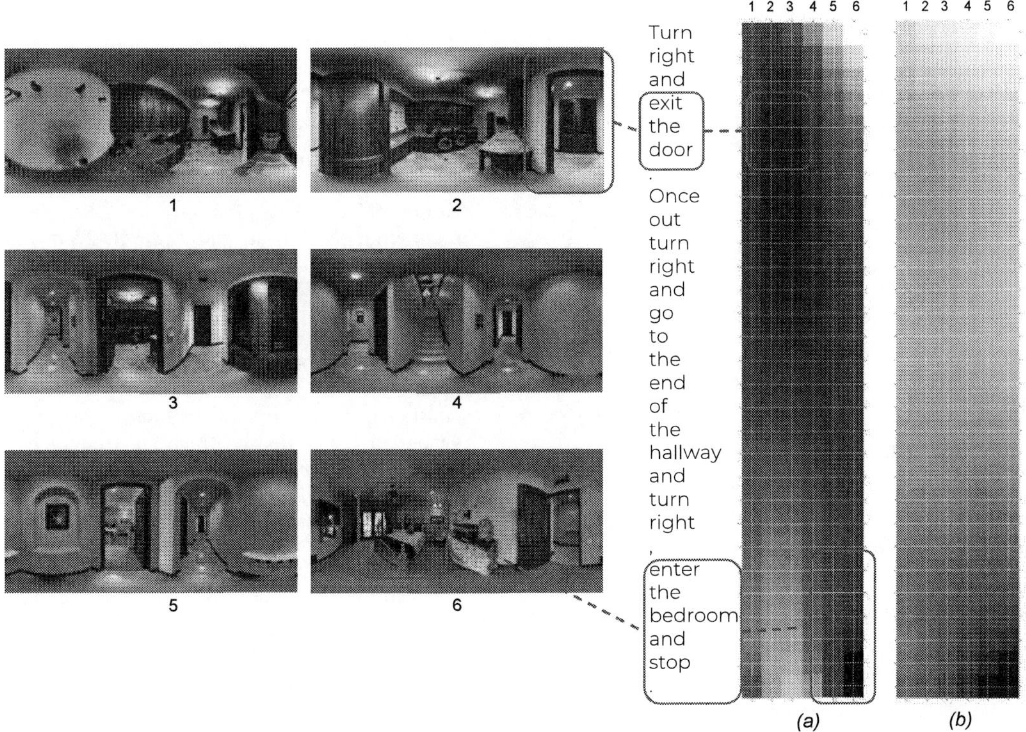

Figure 3: Alignment matrix (Eq.5) for discriminator model trained (a) with curriculum learning on the dataset containing PS, PR, RW negatives (b) without curriculum learning on the dataset with PS negatives only. Note that darker means higher alignment.

forcing. The agent benefits from the multi-modal alignment learned by the discriminator and outperforms the benchmark on the *Validation Unseen* set, as shown in Table 5. This is the condition that best informs how well the agent generalizes. Nevertheless, performance drops on *Validation Seen*, so further experimentation will hopefully lead to improvements on both.

4.3 Visualizing Discriminator Alignment

We plot the alignment matrix A (Eq.5) from the discriminator for a given instruction-path pair to try to better understand how well the model learns to align the two modalities as hypothesized. As a comparison point, we also plot the alignment matrix for a model trained on the dataset with PS negatives only. As discussed before, it is expected that the discriminator trained on the dataset containing only PS negatives tends to exploit easy-to-find patterns in negatives and make predictions without carefully attending to full instruction-path sequence.

Fig.3 shows the difference between multi-modal alignment for the two models. While there is no clear alignment between the two sequences for the model trained with PS negatives only (except maybe towards the end of sequences, as expected), there is a visible diagonal pattern in the alignment for the best discriminator. In fact, there is appreciable alignment at the correct positions in the two sequences, e.g., the phrase *exit the door* aligns with the image(s) in the path containing the object *door*, and similarly for the phrase *enter the bedroom*.

5 Related Work

The release of Room-to-Room (R2R for short) dataset (Anderson et al., 2018b) has sparked research interest in multi-modal understanding. The dataset presents a unique challenge as it not only substitutes virtual environments (e.g., MacMahon et al. (2006)) with photo-realistic environments but also describes the paths in the environment using human-annotated instructions (as opposed to formulaic instructions provided by mapping applications e.g., Cirik et al. (2018)). A number of methods (Anderson et al., 2018b; Fried et al., 2018; Wang et al., 2018a; Ma et al., 2019a; Wang et al., 2018b; Ma et al., 2019b) have been proposed recently to solve the navigation task described in R2R dataset. All these methods build models for agents that learn to navigate in R2R environment and are trained on the entire R2R dataset as well as the augmented dataset introduced by Fried et al. (2018) which is generated by a speaker model trained on human-annotated instructions.

Our work is inspired by the idea of Generative Adversarial Nets (Goodfellow et al., 2014), which use a discriminative model to discriminate real and fake distribution from generative model. We propose models that learn to discriminate between high-quality instruction-path pairs from lower quality pairs. This discriminative task becomes important for VLN challenges as the data is usually limited in such domains and data augmentation is a common trick used to overcome the shortage of available human-annotated instruction-path pairs. While all experiments in this work focus on R2R dataset, same ideas can easily be extended to improve navigation agents for other datasets like Touchdown (Chen et al., 2018).

6 Conclusion

We show that the discriminator model is capable of differentiating high-quality examples from low-quality ones in machine-generated augmentation to VLN datasets. The discriminator when trained with *alignment based* similarity score on cheaply mined negative paths learns to align similar concepts in the two modalities. The navigation agent when initialized with the discriminator generalizes to instruction-path pairs from previously unseen environments and outperforms the benchmark.

For future work, the discriminator can be used in conjunction with generative models producing extensions to human-labeled data, where it can filter out low-quality augmented data during generation as well as act as a reward signal to incentivize generative model to generate higher quality data. The multi-modal alignment learned by the discriminator can be used to segment the instruction-path pair into several shorter instruction-path pairs which can then be used for creating a curriculum of easy to hard tasks for the navigation agent to learn on. It is worth noting that the trained discriminator model is general enough to be useful for any downstream task which can benefit from such multi-modal align-ment measure and not limited to VLN task that we use in this work.

References

Peter Anderson, Angel Chang, Devendra Singh Chaplot, Alexey Dosovitskiy, Saurabh Gupta, Vladlen Koltun, Jana Kosecka, Jitendra Malik, Roozbeh Mottaghi, Manolis Savva, and Amir R. Zamir. 2018a. On evaluation of embodied navigation agents. *arXiv preprint arXiv:1807.06757*.

Peter Anderson, Qi Wu, Damien Teney, Jake Bruce, Mark Johnson, Niko Sünderhauf, Ian Reid, Stephen Gould, and Anton van den Hengel. 2018b. Vision-and-language navigation: Interpreting visually-grounded navigation instructions in real environments. In *Proceedings of the IEEE Conference on Computer Vision and Pattern Recognition (CVPR)*.

Yoshua Bengio, Jérôme Louradour, Ronan Collobert, and Jason Weston. 2009. Curriculum learning. In *Proceedings of the 26th Annual International Conference on Machine Learning*, ICML '09, pages 41–48, New York, NY, USA. ACM.

Howard Chen, Alane Suhr, Dipendra Kumar Misra, Noah Snavely, and Yoav Artzi. 2018. Touchdown: Natural language navigation and spatial reasoning in visual street environments. *CoRR*, abs/1811.12354.

Xinlei Chen, Hao Fang, Tsung-Yi Lin, Ramakrishna Vedantam, Saurabh Gupta, Piotr Dollár, and C. Lawrence Zitnick. 2015. Microsoft COCO captions: Data collection and evaluation server. *CoRR*, abs/1504.00325.

Volkan Cirik, Yuan Zhang, and Jason Baldridge. 2018. Following formulaic map instructions in a street simulation environment. In *2018 NeurIPS Workshop on Visually Grounded Interaction and Language*.

Daniel Fried, Ronghang Hu, Volkan Cirik, Anna Rohrbach, Jacob Andreas, Louis-Philippe Morency, Taylor Berg-Kirkpatrick, Kate Saenko, Dan Klein, and Trevor Darrell. 2018. Speaker-follower models for vision-and-language navigation. In *Neural Information Processing Systems (NeurIPS)*.

Ross B. Girshick, Jeff Donahue, Trevor Darrell, and Jitendra Malik. 2014. Rich feature hierarchies for accurate object detection and semantic segmentation. In *2014 IEEE Conference on Computer Vision and Pattern Recognition, CVPR 2014, Columbus, OH, USA, June 23-28, 2014*, pages 580–587.

Ian Goodfellow, Jean Pouget-Abadie, Mehdi Mirza, Bing Xu, David Warde-Farley, Sherjil Ozair, Aaron Courville, and Yoshua Bengio. 2014. Generative adversarial nets. In Z. Ghahramani, M. Welling, C. Cortes, N. D. Lawrence, and K. Q. Weinberger, editors, *Advances in Neural Information Processing Systems 27*, pages 2672–2680. Curran Associates, Inc.

David Harwath, Adrià Recasens, Dídac Surís, Galen Chuang, Antonio Torralba, and James R. Glass. 2018. Jointly discovering visual objects and spoken words from raw sensory input. In *Computer Vision - ECCV 2018 - 15th European Conference, Munich, Germany, September 8-14, 2018, Proceedings, Part VI*, pages 659–677.

Karl Moritz Hermann, Mateusz Malinowski, Piotr Mirowski, Andras Banki-Horvath, and Raia Hadsell Keith Anderson. 2019. Learning to follow directions in street view. *CoRR*, abs/1903.00401.

Chih-Yao Ma, Jiasen Lu, Zuxuan Wu, Ghassan Al-Regib, Zsolt Kira, Richard Socher, and Caiming Xiong. 2019a. Self-monitoring navigation agent via auxiliary progress estimation. In *Proceedings of the International Conference on Learning Representations (ICLR)*.

Chih-Yao Ma, Zuxuan Wu, Ghassan AlRegib, Caiming Xiong, and Zsolt Kira. 2019b. The regretful agent: Heuristic-aided navigation through progress estimation.

Matt MacMahon, Brian Stankiewicz, and Benjamin Kuipers. 2006. Walk the talk: Connecting language, knowledge, action in route instructions. In *In Proc. of the Nat. Conf. on Artificial Intelligence (AAAI*, pages 1475–1482.

Jeffrey Pennington, Richard Socher, and Christopher Manning. 2014. Glove: Global vectors for word representation. In *Proceedings of the 2014 Conference on Empirical Methods in Natural Language Processing (EMNLP)*, pages 1532–1543. Association for Computational Linguistics.

Mike Schuster and Kuldip K. Paliwal. 1997. Bidirectional recurrent neural networks. *IEEE Trans. Signal Processing*, 45:2673–2681.

Oriol Vinyals, Alexander Toshev, Samy Bengio, and Dumitru Erhan. 2017. Show and tell: Lessons learned from the 2015 MSCOCO image captioning challenge. *IEEE Trans. Pattern Anal. Mach. Intell.*, 39(4):652–663.

Harm de Vries, Kurt Shuster, Dhruv Batra, Devi Parikh, Jason Weston, and Douwe Kiela. 2018. Talk the walk: Navigating new york city through grounded dialogue. *CoRR*, abs/1807.03367.

Jiang Wang, Yang Song, Thomas Leung, Chuck Rosenberg, Jingbin Wang, James Philbin, Bo Chen, and Ying Wu. 2014. Learning fine-grained image similarity with deep ranking. In *Proceedings of the 2014 IEEE Conference on Computer Vision and Pattern Recognition*, CVPR '14, pages 1386–1393, Washington, DC, USA. IEEE Computer Society.

Xin Wang, Qiuyuan Huang, Asli Çelikyilmaz, Jianfeng Gao, Dinghan Shen, Yuan-Fang Wang, William Yang Wang, and Lei Zhang. 2018a. Reinforced cross-modal matching and self-supervised imitation learning for vision-language navigation. *CoRR*, abs/1811.10092.

Xin Wang, Wenhan Xiong, Hongmin Wang, and William Yang Wang. 2018b. Look before you leap: Bridging model-free and model-based reinforcement learning for planned-ahead vision-and-language navigation. In *Computer Vision – ECCV 2018*, pages 38–55, Cham. Springer International Publishing.

Haonan Yu and Jeffrey Mark Siskind. 2013. Grounded language learning from video described with sentences. In *Proceedings of the 51st Annual Meeting of the Association for Computational Linguistics, ACL 2013, 4-9 August 2013, Sofia, Bulgaria, Volume 1: Long Papers*, pages 53–63.

Matthew D. Zeiler and Rob Fergus. 2014. Visualizing and understanding convolutional networks. In *Computer Vision - ECCV 2014 - 13th European Conference, Zurich, Switzerland, September 6-12, 2014, Proceedings, Part I*, pages 818–833.

Semantic Spatial Representation: a unique representation of an environment based on an ontology for robotic applications

Guillaume Sarthou
LAAS-CNRS, Université
de Toulouse, CNRS,
Toulouse, France
gsarthou@laas.fr

Aurélie Clodic
LAAS-CNRS, Université
de Toulouse, CNRS,
Toulouse, France
aclodic@laas.fr

Rachid Alami
LAAS-CNRS, Université
de Toulouse, CNRS,
Toulouse, France
alami@laas.fr

Abstract

It is important, for human-robot interaction, to endow the robot with the knowledge necessary to understand human needs and to be able to respond to them. We present a formalized and unified representation for indoor environments using an ontology devised for a route description task in which a robot must provide explanations to a person. We show that this representation can be used to choose a route to explain to a human as well as to verbalize it using a route perspective. Based on ontology, this representation has a strong possibility of evolution to adapt to many other applications. With it, we get the semantics of the environment elements while keeping a description of the known connectivity of the environment. This representation and the illustration algorithms, to find and verbalize a route, have been tested in two environments of different scales.

1 Introduction

Asking one's way, when one does not know exactly where one's destination is, is something we all did. Just as we have all responded to such a request from a lost person. This is the heart of the road description task. This task, which seems so natural to us in a human-human context, requires in fact a set of knowledge (e.g. on the place, on the possible points of reference in this particular environment) and "good practices" (e.g. to give a path easy to follow if possible) that need to be modeled if we want to implement it on a robot. This paper presents a robotics application of our system but it would be possible to use it in other applications such as virtual agent.

This route description task is an interesting application case through the variety of the needed information (e.g. type of elements, place topology, names of the elements in natural language). It has been well studied in the field of human-robot interaction. Robot guides have already been deployed in shopping centers (Okuno et al., 2009), museums (Burgard et al., 1998; Clodic et al., 2006; Siegwart et al., 2003) or airports (Triebel et al., 2016). However, using metrical (Thrun, 2008), topological (Morales Saiki et al., 2011), semantic representations, or trying to mix them together (Satake et al., 2015b) (Chrastil and Warren, 2014) (Zender et al., 2008), it is difficult to have a uniform way to represent the environment. In addition, it is difficult to have a representation which allows to calculate a route and at the same time to express it to the human with whom the robot interacts, because this requires data of different types. Our aim is to propose a single and standardized representation of an environment which can be used to choose the appropriate route to be explained to the human and at the same time to verbalize it using a route perspective. The purpose of this paper is not to be applied to a guiding task in which a mobile robot accompanies the human to his final destination but to explain to a human how to reach it. Consequently, we will not talk here about a metrical representation like the one that can be used to navigate in the environment (Thrun, 2008) or to negotiate it use in (Skarzynski et al., 2017).

Route perspective means essentially to navigate mentally in order to verbalize the path to follow but also to facilitate understanding and memorizing instructions. The route perspective opposes the survey perspective which is a top view with landmarks and paths printed on a map. Morales et al. (Morales et al., 2015) indicate that naming parts of a geometric map does not leave the opportunity to compute such perspective. As in (Satake et al., 2015a), we have chosen to develop our representation with an ontology as it allows to reason about the meaning of words and thus improve the understanding of human demands. In addition, we propose a way to merge the topological representation into the semantic representation

Proceedings of SpLU-RoboNLP, pages 50–60
Minneapolis, MN, June 6, 2019. ©2019 Association for Computational Linguistics

(the ontology) to get the meaning of the environment elements while keeping a description of the connectivity of the elements of the environment. We propose to name it semantic spatial representation (SSR). With this, we are able to develop the two features presented in (Mallot and Basten, 2009) for the route description task, which consist of selecting a sequence of places leading to the objective and managing the declarative knowledge to choose the right action to explain at each point of the sequence. Based on the principles of topological description, although represented semantically, we are able to compute multiple routes and new detours for the same objective in contrast with a route knowledge, which maps a predefined route to a given request. Thanks to this capacity and to the semantic knowledge of the environments available in the representation, it is also possible to provide the most relevant route to a user according to his preferences and capabilities. A basic example would be that we will never recommend a path with stairs to a mobility impaired person. More than the extension of the spatial semantic hierarchy (SSH) (Kuipers, 2000) allowing the representation of the environment, we present here an algorithm to choose the routes and another one to generate an explanation sentence. Both algorithms are based solely on the knowledge provided by the SSR.

Regarding the representation of the environment generally used in order to find an itinerary, we have first to analyse GNSS road navigation systems. In (Liu, 1997) or (Cao and Krumm, 2009), we find the same principle of a topological network representing the roads with semantic information attached to each of them. This type of representation seems logical regarding the performance required for such systems operating in very large areas. However, GNSS road navigation systems must respond only to this unique task of finding a path when a robot is expected to be able to answer to various tasks. This is why we have developed and implemented a representation that can be used more widely while still allowing the search for routes.

This paper focuses on the presentation of the SSR and on its usability for the route description task. For now, all the ontologies used to test the SSR have been made by hand. However, many recent research work leads to automatically generate a topological representation of an environment from geometric measurements (e.g. Region Adjacency Graphs (Kuipers et al., 2004), Cell and Portal Graphs (Lefebvre and Hornus, 2003) or hierarchical models (Lorenz et al., 2006), or from natural language (Hemachandra et al., 2014)). We have not done it yet, but our system could benefit from this work to generate a representation of an environment using SSR, which would solve the complexity of creating such a representation by hand.

In order to present our work, we will follow the three cognitive operations needed to generate a spatial discourse (Denis, 1997): (section 2) an activation of an internal representation of the environment; (sections 3 and 4) the planning of a route in the mental representation made previously; (section 5) the formulation of the procedure that the user must perform to achieve the objective. The SSR and the algorithms demonstrating its usability have been tested in two environments of different scales: an emulated mall in our laboratory and a real mall. Results are presented in section 6 for the two environments.

2 Environment representation: SSR

In cognitive psychology, Semantic memory refers to the encyclopedic knowledge of words associated to their meanings. Some authors have proposed a model of this semantic memory as being a semantic network in which each concept is linked to others by properties and have designed a computer-implemented model (Collins and Quillian, 1969). This initial model has since been formalized as an ontology (Berners-Lee et al., 2001) and is already widely used in the semantic web.

This model is already used in robotics to obtain a detailed representation of the environment in which robots operate. For example, (Satake et al., 2015a) and (Beetz et al., 2018) use an ontology to represent knowledge about the types of items such as the types of shops (restaurant, fashion store, for example) or the properties of items such as the stores where they are sold.

(Kuipers, 2000) introduced the 'topological level' with SSH (spatial semantic hierarchy) which defines a place, a path and a region and defined several relationships between them. Ontologies are constructed using triplets where two concepts are linked by a property (e.g $property(concept1, concept2)$), however the Kuipers SSH does not allow such representation due to the use of some quadruplets (e.g

$along(view, path, dir)$) in addition to triplets. To overcome this limitation, we propose a formalisation, that we call Semantic Spatial Representation (SSR) to represent an environment with ontologies (i.e. using triplets).

In this section we present the minimal ontology that constitutes the SSR but it can be extended to represent the knowledge of the types and the properties of the elements while preserving the first use of this model.

2.1 Classes

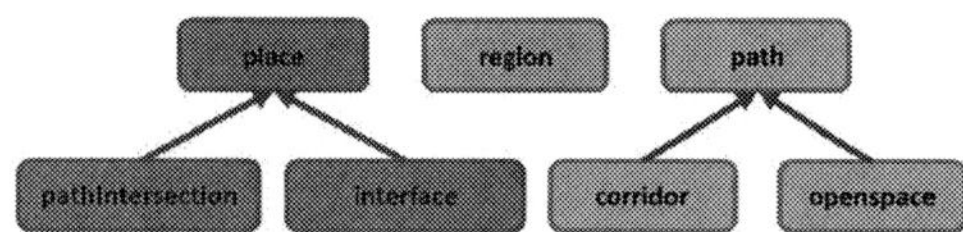

Figure 1: Classes for a representation of the topology of an indoor environment in a semantic description.

Region: It represents a two-dimensional area that is a subset of the overall environment. A description of the environment must include at least one region representing the entire environment. Regions are used to reduce the complexity of the routes computation, so we recommend to use several region especially for large scale areas. A basic use of the regions is for multi-storey buildings where each floor should be more naturally considered as a region. Regions can be described as being nested.

Path: It is a one dimensional element along which it is possible to move. A path must have a direction.

- *Corridor*: It represents a kind of path with a beginning and an end, for which beginning and end are distinct. The arbitrary direction chosen for a corridor defines the position of its beginning and end. This defines also the right and left of the corridor.

- *Openspace*: It is a kind of path which does not have any begin or end. It can be viewed as a "potato-shaped" describing the outline of an open space. It materializes the possibility of turning the gaze around the room and the fact of not having to go through a defined path to reach one of its points.

Place: It represents a point of zero dimension that can represent a physical or symbolic element. It can be extended to represent stores and landmarks in the example of a shopping center.

- *Path intersection*: It represents the connection between only two paths and thus a waypoint to go from one path to another. In the case of a crossing between three paths, three intersections would therefore be described.

- *Interface*: It represents the connection between only two regions and thus a waypoint to move from one region to another. It can be physical, like a door or a staircase, or symbolic like a passage.

The distinction between paths and places is related to the differences between the types of rooms made by (Andresen et al., 2016) where some are used to circulate (corridors) while others have an explicit use to the exclusion of traffic (place).

2.2 Properties

Properties are used to express topological relationships such as connections between paths and places or the order of places along a path. All the properties presented here can be extended with their inverse (e.g. $isIn$ and $hasIn$) for a more expressive model and thus easier handling.

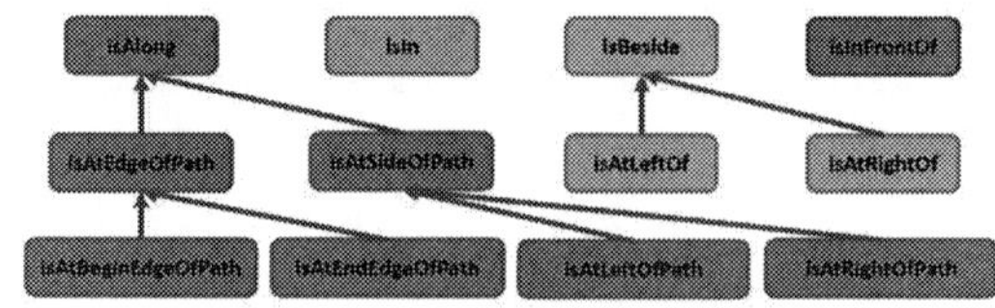

Figure 2: Properties for a representation of the topology of an indoor environment in a semantic description.

$isIn(path/place, region)$: *path* or *place* is in *region*.
$isAlong(place, path)$: *place* is along *path*.

- $isAlong(place, openspace)$: For open spaces, since there is no beginning or end, places are only defined as being along an open space.

- $isAlong(place, corridor)$: For corridors, the specific properties $isAtBeginEdgeOfPath$, $isAtEndEdgeOfPath$, $isAtLeftOfPath$, $isAtRightOfPath$ must be used. The choice of these properties is made with the arbitrary direction defined by positioning itself at its beginning and by traversing it towards its end.

$isBeside(place1, place2)$: $place1$ is beside $place2$. Specified properties $isAtLeftOf$ and $isAtRightOf$ must be used to express the order of places. The choice of these properties is made by positioning themselves at the place and facing the path along the place.

$isInfrontOf(place1, place2)$: $place1$ is in front of $place2$. This property does not need to be applied to all the places described. The more it is used, the more the verbalization of the itinerary will be easy. It is important to always define a place in front of an intersection to be able to determine if the guided human will have to go left or right in some cases. If there is no described place in front of an intersection, we can use a $emptyPlace$ class that would inherit the $place$ class.

The following axioms reduce the complexity of the description of the environment. The logical relations will therefore be solved by the ontology reasoner.

- $isAtLeftOf(place1, place2)$ $\leftrightarrow$ $isAtRightOf(place2, place1)$

- $isInfrontOf(place1, place2)$ $\leftrightarrow$ $isInfrontOf(place2, place1)$

- $isAlong(place, path)$ $\wedge$ $isIn(path, region) \rightarrow isIn(place, region)$

3 Computing routes

At this point, we have built an internal representation of the environment using the Semantic Spatial Representation (SSR). We illustrate how this representation can be used to compute the possible routes from one place to another. Even if the length of a route is taken into account in the choice, the complexity of the description is an important criterion (Morales et al., 2015). When someone asks his way, the guide will not necessarily try to give him the shortest way. His main goal is to make sure the person reach her goal. In the example of Figure 3, even if the red route is little longer than blue route, he will certainly propose it instead. Every intersection or change of direction is a risk for the guided person to make a mistake and thus get lost.

In this section, the goal is to provide multiple routes so that we can allow to choose the best route based on the person preferences. The possibility

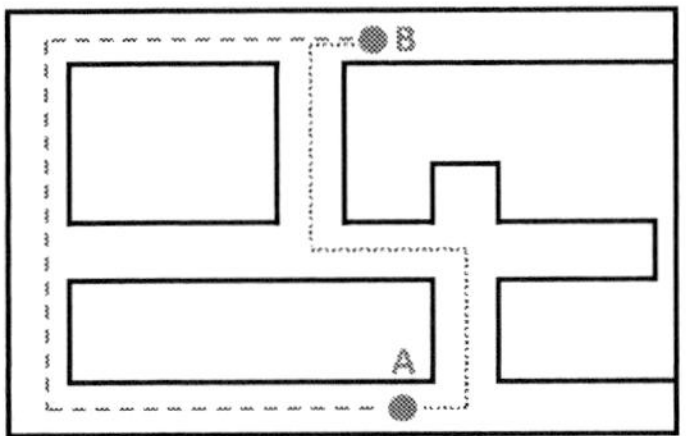

Figure 3: Comparison of two routes in terms of complexity and length. The blue route (. . .) is the shortest but is complex to explain while the red (- - -) is simpler although a bit longer.

of making this choice using the SSR will be presented in section 4. In order to reduce the complexity of the search, especially for large scale environments, we propose to work at two levels:

- **First :** Region level: considers only areas and passages such as doors, stairs or elevators.

- **Then :** Place level: provides complete routes description including paths and intersections within regions.

3.1 Region level

In large-scale environments such as multi-storey buildings, routes computation can lead to combinatorial explosion. Exploration at the region level decreases this effect by conducting a first high-level exploration. In Figure 4 we can see that the exploration of paths of regions 4 and 5 is useless because these regions do not lead to the region containing the goal. This exploration uses only the regions and interface elements described in section 2.

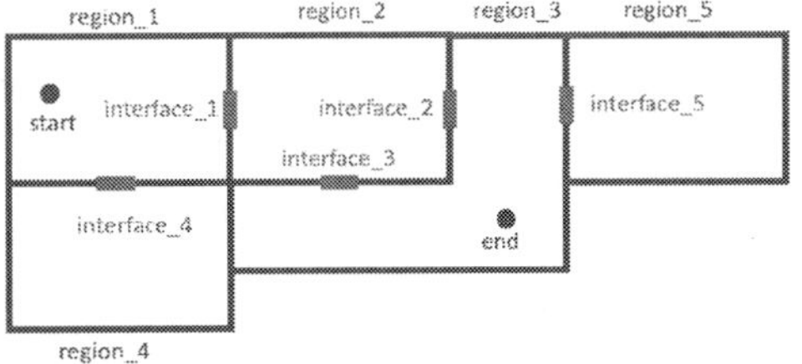

Figure 4: Representation of an environment at the regional level.

Each interface is connected to regions thanks to the $isIn$ property. With this property, a route finding algorithm, based on the breadth-first search, makes possible to find the shortest routes connecting two regions by using the semantic knowledge.

By including the knowledge base exploration directly inside the search algorithms, it it not necessary to extract a topological graph with nodes and arcs. It is carried out within the search algorithm without preprocessing.

This algorithm applied to the example presented in Figure 4 gives the tree of Figure 5. The final routes found by the algorithm are :

- $region\_1 - interface\_1 - region\_2 - interface\_2 - region\_3$

- $region\_1 - interface\_1 - region\_2 - interface\_3 - region\_3$

Region 5 has never been explored and region 4 is not present in the final result. However, both solutions with interfaces 2 and 3 have been taken into account. This type of results makes possible to quickly eliminate unnecessary solutions and thus reduces the complexity for a more detailed search in a second time. This technique is similar to what is done for GNSS road navigation systems where the main roads are studied upstream of secondary roads with pyramidal (or hierarchical) route structure (Bovy and Stern, 1990).

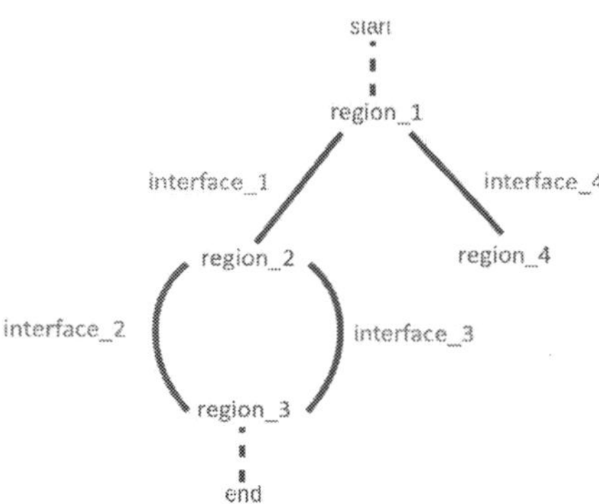

Figure 5: Exploration graph resulting from region-level search (sec.3.1) and the aggregation of start and end places (sec.3.2) .

3.2 Place level

Place-level search is based on the Region-level search results with the aggregation of start and end places, so the format changes from $region - place - region - ... - region$ to a $place - region - place - ... - place$.

Place-level search works from one place to another through a single region. We have therefore divided the previous solutions to meet this constraint. This step aims to reduce complexity again.

Indeed, if several routes pass through the same region with the same places of departure and arrival, the inner route can be calculated once and for all. In our example, the division gives five sub-routes instead of six:

- $start - region\_1 - interface\_1$

- $interface\_1 - region\_2 - interface\_2$

- $interface\_2 - region\_3 - end$

- $interface\_1 - region\_2 - interface\_3$

- $interface\_3 - region\_3 - end$

The place-level algorithm aims to replace each sub-route region with a succession of paths and intersections. It works on the same principle as the previous search algorithm using the $isAlong$ property instead of the $isIn$ property. To improve performance, we use moving forward for the breadth-first search. It stops the exploration of the routes passing through a path already used in previous route calculation steps. In addition, it prevents loops.

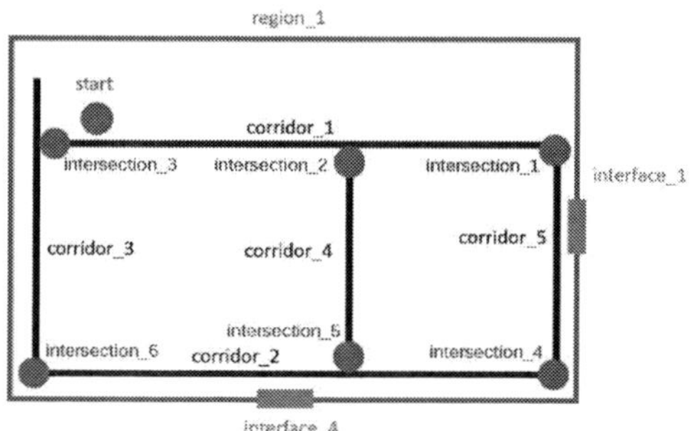

Figure 6: Representation of corridors and intersections in region 1 from the example 4

Taking the example of Figure 4 and focusing on region 1, we can solve the sub-route $start - region\_1 - interface\_1$. Region 1 is represented with its corridors and intersections in Figure 6. By applying the algorithm at the place level, we have the solution $start - corridor\_1 - intersection\_1 - corridor\_5 - interface\_1$. By doing the same for each sub-route, we can then recompose the global routes and give the set detailed of routes from start to end.

4 Choosing the best route

Since the SRR is based on an ontology, we can have the meaning of each element of the environment and we can attach additional information to them as features. Now that we have found several routes to the same goal, we want to select one based on different criteria. This selection of routes is independent of the previous section and a variety of cost functions can be implemented based on specific application needs. In the following subsection, we present an example of cost function using SSR and designed for robot guide to be deployed by the European project MuMMER (Foster et al., 2016).

4.1 Example of cost function

As mentioned in (Morales et al., 2015), the complexity of the route to explain to a human, which is the number of steps in a route, is the most important criterion in choosing the route. A cost function taking into account only the complexity of a route R in the environmental context M would be $f(R, M)$ with N being the number of steps of R.

$$f(R, M) = N \qquad (1)$$

However, to find a good itinerary, it is important to take into account the preferences and capabilities of the guided person. An easy example is that we will never indicate a route with stairs to a person with reduced mobility. Using an ontology and therefore the possibility of describing the properties of elements of the environment, we add the property $hasCost$ which associates an element with a criterion. Criteria rated σ_i are: saliency, accessibility, comfort, security, ease of explanation and speed. Other criteria could easily be added through the use of ontology according to the specific needs of the environment. All these criteria and their antonyms can be applied to each element n. The preferences of a person P are costs related to the criterion σ_i noted φ_{σ_i}. This represents the sensitivity of P to the σ_i criterion. The cost function becomes $f(P, R, M)$ to take into account the preference of person P.

$$f(P, R, M) = N \times \prod_{n=0}^{N} [\prod_i (\sigma_{in} \times \varphi_{\sigma_i})] \qquad (2)$$

Because we focused only on the complexity of the route explanation and the characteristics of the elements of the environment, in the presented cost function 2, the distances are not taken into account. This information could be added by working with a metric representation. Another possibility that can be explored is to add some of the metric knowledge, such as the length of the corridor, into the semantic representation of the environment to preserve the working principle of a unique representation of the environment in this route description task.

5 Explanation generation

This section describes the third cognitive operation of (Denis, 1997) to generate a spatial discourse: the formulation of the procedure. As (Kopp et al., 2007), we define a route description as a set of route segments, each connecting two important points and explained in a chronological way. As (Tversky and Lee, 1999), we add to each route segment a triplet: orientation, action, and landmark to enable its explanation. The division into segments corresponds to all paths, with their entry and exit points, provided by our planning part. However, the semantic representation (SSR) used to plan the route is not directly usable to generate the formulation of the procedure. With the current representation, the orientation and action are too complex to extract (given that they depend on the direction by which the person arrives). It is however possible to interpret the semantic representation in relation to the estimated future position of the human. This interpretation is what we call the internal representation. This internal representation is composed of several sub-representations each representing a path of the global environment. Each segment of the route is represented independently of the others. For open space, we generate an ordered array of all locations *along* it. For the corridors, we generate four ordered arrays to represent the *left*, the *right*, the *beginedge* and the *endedge* of the corridors. These information can be found in the ontology with the properties *isAlong*, *isAtLeftOfPath*, and so on. To order the places in each array, we use the properties *isAtLeftOf* and *isAtRightOf* also present in the ontology. This internal representation can be displayed and gives Figure 7 for the corridor_1 of region_1 from the example 4. The *isInfrontOf* property is used to generate better placements.

Once we have an internal representation of each segment, we can determine the procedure that the user must perform. (Kopp et al., 2007) mention that an action, a reorientation, a progression or a positioning must be carried out at the end of each segment. The end of one

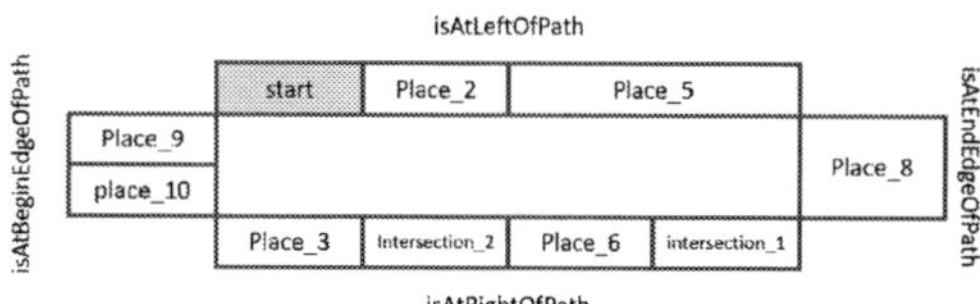
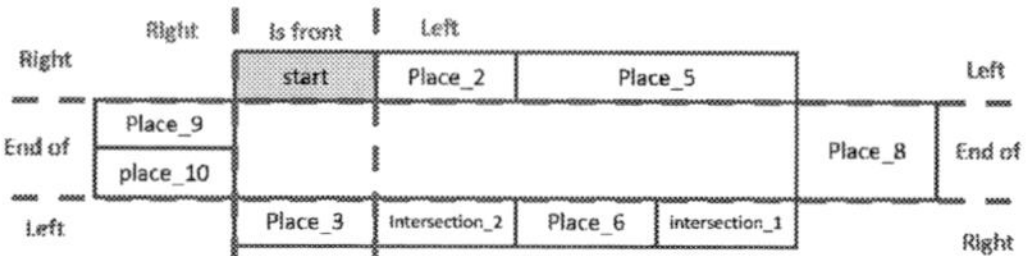

Figure 7: Internal representation of a the corridor_1 of region_1 from the example 6, extracted from the semantic representation.

Figure 8: Resolution of directions and directions with an entry in the hallway by the gray square "start".

segment being the beginning of the next, we choose to determine the actions at the beginning of each segment (which corresponds more to our internal representation). It allows to work on one path at a time. This rule is formalized as *"choosing action A_i at place P_j will lead to place P_k"* by (Mallot and Basten, 2009). This determination of actions can be made with our internal representation, as shown in Figure 8 where P_j is the gray place and P_k can be one of the other. The red information at the top gives the *"turn right"* action with P_k being *Place_9* or *Place_10*, *"go in front of you"* with P_k being *Place_3* and *"turn left"* for the other places. The blue information on the sides gives the orientation of the sub-goal place P_k taking into account the previous reorientation. With this orientation information we can give an explanation of the form *"take the corridor at your right"* where the action is determined by the type of P_k and the side given by the orientation information. On the example of corridor_1, to go from *start* to *intersection_1*, the full sentence will be *turn left then take the corridor at your right*. Moreover, by taking into account the orientation of the guided person after an action we allow to provide directions in the route perspective and so the guided person to perform an imaginary tour of the environment (Kopp et al., 2007).

By working segment by segment in the order given by the route search algorithm, we necessarily generate the explanations with a temporospatial ordering. This criterion is an important point in Allen's best practice in communicating route knowledge (Allen, 2000).

The latest critical information presented by (Tversky and Lee, 1999) is landmark and we have it in our representation. (Tversky and Lee, 1998) noted that more than 90% of the guiding spots on maps and verbal directions contained additional information, which corresponds also to the results of (Denis, 1997) and the Allen's best practice (Allen, 2000). With our internal representation, we provide all the landmarks (corresponding to places because being defined as such) around which action must be taken and we can therefore refer to it to help the guided person. On the previous example, the sentence may be confusing because there are two corridors on the left. We are able to refer to *place_8* which will be on the left or *place_6* which will be on the right by projecting the future position of the human at *intersection_1*. With this new information, the situation can be disambiguated. The full sentence will became *turn left then take the corridor at your right straight after place_6*.

The verbalization software based on the SSR and the principles described above was created based on a human-human study (Belhassein et al., 2017). Among the set of route description sentences, we have identified four types of explanatory components: those corresponding to the beginning of route description, to the end of a route, of progress in the route and the particular case of explanations with one step. These four types are only dependent of the position of the segment to be included in the global explanation procedure. For each type, we have identified various sub-types depending on the actions to be performed or the location of actions. For example, for the types of end-of-procedure sentences, we distinguish those where the end goal will be on the side, in front or at the current future position. In total, we have identified 15 sub-types. Each component of the explanation sentence has been classified into one of these sub-types. We want to be able to propose different ways to express the same things so the system does not have only one way to express the very same information. To represent similar sentences and to be able to generate sentences with variations, we have grouped sentences with close lexical structures. Each sentence is then represented with its variations as follows: ["you will "], ["see

", "find "],["it ", "/X "], ["on "], ["your ", "the "], ["/D "],["side ", "when you walk ", ""]. When using a sentence, the variations are randomly chosen with a uniform distribution. We can notice in the previous example the use of variables such that X which corresponds to the name of an element of the environment and D to a direction. We also used the Y variables for a reference points and DY for a reference point directions. If a sentence requires a variable that we have not been able to extract from our internal representation, then another sentence with the same meaning or another variation of the sentence that does not require the variable is chosen.

6 Applications

The SSR was first applied in an emulated mall to develop the algorithms [1], but we also tested it in a real mall to study its applicability in a larger environment. Table 1 indicates the number of elements described in both environments. The number of places does not correspond only to the sum of the shops, interfaces and intersections because much more elements have been described, such as ATMs, restrooms or carts location.

	emulated	real
place	83	249
shop	19	135
interface	11	18
path intersection	10	52
path	11	42
region	5	4

Table 1: Number of elements described in the emulated and real environment.

Table 2 presents the CPU time needed for the routes computation and cost function algorithms for several cases, applied to the real environment representation. Even though specialized algorithms that work with a specific representation of the environment may be faster than ours, we can see here that they are acceptable in the context of a human robot interaction and especially in a route description task to both compute the path and verbalize it. Indeed, by providing semantic, topological and spatial knowledge within a single representation it can be used by several algorithms usually requiring different representations.

We can also see the use of the regions to reduce the computation times with the two cases where three routes were found, one of the cases crossing one region and the other two.

Number of routes found	Number of regions crossed	Number of paths used	Path finding execution time (ms)
1	1	1	< 10
1	1	3	$[20, 25]$
3	1	9	$[50, 55]$
3	2	12	$[30, 35]$
16	2	75	$[160, 170]$
20	2	129	$[180, 190]$

Table 2: CPU time (min-max interval) time for computing routes in a big shopping mall description. Each row refers to a single request that can provide multiple routes to the goal.

To show the usability of the internal representation extracted from the SSR in the verbalization of the route, we have developed a software [2] that is able to verbalize the route found by our semantic planner. In examples of the sentences synthesized by this software (Table 3), we can see that for the same goal, it is possible to use different points of reference and to position them with respect to another element in the environment. All directions shown in the A and B examples take into account the future position of the guided human and provide indications from the perspective of the route.

Goal	Sentence
Y	You see there **Y**.
Y	It's on the right side of **Z**.
Y	It's on the left side of **X**.
A	Go through the door. Take the stairs at your left and turn left. Go almost at the very end of the corridor and, turn left at the door. After that you will see **A** on your right.
B	Go straight down this aisle. Then, walk to the very end of the corridor and it's on the left there.

Table 3: Sentences generated by a software using the internal representation extracted from the SSR.

The applications presented previously have not

only been tested as such, but have been integrated into a global robotic architecture and deploy in a mall center [3] as shown in figure 9. This integration shows that the results obtained by algorithms working with a single semantic representation of an environment are usable and are relevant in a more global task.

Figure 9: Robot describing a route to a human in a mall using the SSR and the associated algorithms. The sentence in green is the explanation of the route verbalized by the robot from the SSR representation: "just go down the corridor and then go almost at the very end of this corridor and it's on the left when you walk".

7 Conclusions

We have proposed an environment model that suitable to find routes, to choose one and ro be able to verbalize it using a single representation. The key contribution of our work is the **semantic spatial representation (SSR)**, a formalization of how to describe an environment such as a large and complex public space mall using an ontology. We have also presented results about the use of our system by a robot that provides route guiding to humans in a shopping mall.

To benefit from our system, it could be interesting to integrate this representation and the corresponding algorithms to a dialog system (Papaioannou et al., 2018) in order to exploit more deeply its capacities. An interesting usage of this system already possible but not yet exploited because of the need of a dialog system, would be to use the guided person previous knowledge to choose a route and/or to generate an explanation (*"If you know the place_2, from this one ..."*). In the same vein, it would be possible to link it with a system such as Shared Visual Perspective Planner (Waldhart et al., 2018) to begin explaining the route from a visible point. This would reduce the length of the

---

[3]https://cloud.laas.fr/index.php/s/CJcPWmMU7TZGQJB

explanations and thus ensure a better understanding of the itinerary for the guided person. Another improvement would be to use an ontology to ground the interaction (Lemaignan et al., 2012) as part of the route description task.

At this stage, only the topological representation has been integrated into the semantic representation. This is a good first step in working with a single representation that is easier to evolve and ensure consistency of knowledge. Future work would involve the integration of metric information, and thus geometric representation.

Acknowledgments

This work has been supported by the European Unions Horizon 2020 research and innovation programme under grant agreement No. 688147 (MuMMER project).

References

Gary L. Allen. 2000. Principles and practices for communicating route knowledge. *Applied cognitive psychology*, 14(4):333359.

Erik Andresen, David Haensel, Mohcine Chraibi, and Armin Seyfried. 2016. Wayfinding and Cognitive Maps forPedestrian Models. In *Traffic and Granular Flow '15*, pages 249–256. Springer International Publishing.

Michael Beetz, Daniel Beler, Andrei Haidu, Mihai Pomarlan, and Asil Kaan Bozcuog. 2018. KnowRob 2.0 A 2nd Generation Knowledge Processing Framework for Cognition-enabled Robotic Agents. page 8.

Kathleen Belhassein, Aurélie Clodic, Hélène Cochet, Marketta Niemelä, Päivi Heikkilä, Hanna Lammi, and Antti Tammela. 2017. Human-Human Guidance Study.

Tim Berners-Lee, James Hendler, and Ora Lassila. 2001. The Semantic Web. *Scientific American*, 284(5):34–43.

P. H. Bovy and E. Stern. 1990. *Route Choice: Wayfinding in Transport Networks: Wayfinding in Transport Networks*. Studies in Operational Regional Science. Springer Netherlands.

Wolfram Burgard, Armin B. Cremers, Dieter Fox, Dirk Hähnel, Gerhard Lakemeyer, Dirk Schulz, Walter Steiner, and Sebastian Thrun. 1998. The Museum Tour-Guide Robot RHINO. In *Autonome Mobile Systeme 1998, 14. Fachgespräch, Karlsruhe, 30. November - 1. Dezember 1998*, pages 245–254.

Lili Cao and John Krumm. 2009. From GPS Traces to a Routable Road Map. In *Proceedings of the 17th*

ACM SIGSPATIAL International Conference on Advances in Geographic Information Systems, GIS '09, pages 3–12, New York, NY, USA. ACM.

Elizabeth R. Chrastil and William H. Warren. 2014. From Cognitive Maps to Cognitive Graphs. *PLOS ONE*, 9(11):e112544.

Aurelie Clodic, Sara Fleury, Rachid Alami, and al. 2006. Rackham: An Interactive Robot-Guide. In *The 15th IEEE International Symposium on Robot and Human Interactive Communication, RO-MAN 2006, Hatfield, Herthfordshire, UK, 6-8 September, 2006*, pages 502–509.

Allan M. Collins and M. Ross Quillian. 1969. Retrieval time from semantic memory. *Journal of Verbal Learning and Verbal Behavior*, 8(2):240–247.

Michel Denis. 1997. The description of routes: A cognitive approach to the production of spatial discourse. *Cahiers de Psychologie Cognitive*, 16:409–458.

Mary Ellen Foster, Rachid Alami, Olli Gestranius, Oliver Lemon, Marketta Niemelä, Jean-Marc Odobez, and Amit Kumar Pandey. 2016. The MuMMER Project: Engaging Human-Robot Interaction in Real-World Public Spaces. In *Social Robotics. 8th International Conference, ICSR 2016, Kansas City, MO, USA, November 1-3, 2016 Proceedings*, Lecture Notes in Computer Science, pages 753–763.

Sachithra Hemachandra, Matthew R. Walter, Stefanie Tellex, and Seth Teller. 2014. Learning spatial-semantic representations from natural language descriptions and scene classifications. In *2014 IEEE International Conference on Robotics and Automation (ICRA)*, pages 2623–2630, Hong Kong, China. IEEE.

Stefan Kopp, Paul A. Tepper, Kimberley Ferriman, Kristina Striegnitz, and Justine Cassell. 2007. Trading Spaces: How Humans and Humanoids Use Speech and Gesture to Give Directions. In Toyoaki Nishida, editor, *Wiley Series in Agent Technology*, pages 133–160. John Wiley & Sons, Ltd, Chichester, UK.

B. Kuipers, J. Modayil, P. Beeson, M. MacMahon, and F. Savelli. 2004. Local metrical and global topological maps in the hybrid spatial semantic hierarchy. In *IEEE ICRA, 2004. Proceedings. ICRA '04. 2004*, volume 5, pages 4845–4851 Vol.5.

Benjamin Kuipers. 2000. The Spatial Semantic Hierarchy. *Artificial Intelligence*, 119(1):191–233.

Sylvain Lefebvre and Samuel Hornus. 2003. Automatic Cell-and-portal Decomposition. report, INRIA.

Séverin Lemaignan, Raquel Ros, Emrah Akin Sisbot, Rachid Alami, and Michael Beetz. 2012. Grounding the Interaction: Anchoring Situated Discourse in Everyday Human-Robot Interaction. *International Journal of Social Robotics*, 4(2):pp.181–199. 20 pages.

Bing Liu. 1997. Route finding by using knowledge about the road network. *IEEE Transactions on Systems, Man, and Cybernetics - Part A: Systems and Humans*, 27(4):436–448.

Bernhard Lorenz, Hans Jrgen Ohlbach, and Edgar-Philipp Stoffel. 2006. A Hybrid Spatial Model for Representing Indoor Environments. In *Web and Wireless Geographical Information Systems*, Lecture Notes in Computer Science, pages 102–112. Springer Berlin Heidelberg.

Hanspeter A. Mallot and Kai Basten. 2009. Embodied spatial cognition: Biological and artificial systems. *Image and Vision Computing*, 27(11):1658–1670.

Yoichi Morales, Satoru Såtake, Takayuki Kanda, and Norihiro Hagita. 2015. Building a Model of the Environment from a Route Perspective for Human-Robot Interaction. *International Journal of Social Robotics*, 7(2):165–181.

Luis Yoichi Morales Saiki, Satoru Satake, Takayuki Kanda, and Norihiro Hagita. 2011. Modeling Environments from a Route Perspective. In *Proceedings of the 6th International Conference on Human-robot Interaction*, HRI '11, pages 441–448, New York, NY, USA. ACM. Event-place: Lausanne, Switzerland.

Yusuke Okuno, Takayuki Kanda, Michita Imai, Hiroshi Ishiguro, and Norihiro Hagita. 2009. Providing route directions: design of robot's utterance, gesture, and timing. In *HRI, 2009 4th ACM/IEEE International Conference on*, pages 53–60. IEEE.

Ioannis Papaioannou, Christian Dondrup, and Oliver Lemon. 2018. Human-robot interaction requires more than slot filling - multi-threaded dialogue for collaborative tasks and social conversation. In *FAIM/ISCA Workshop on Artificial Intelligence for Multimodal Human Robot Interaction*. ISCA.

Satoru Satake, Kotaro Hayashi, Keita Nakatani, and Takayuki Kanda. 2015a. Field trial of an information-providing robot in a shopping mall. In *2015 IEEE/RSJ IROS*, pages 1832–1839.

Satoru Satake, Keita Nakatani, Kotaro Hayashi, Takyuki Kanda, and Michita Imai. 2015b. What should we know to develop an information robot? *PeerJ Computer Science*, 1:e8.

Roland Siegwart, Kai Oliver Arras, Samir Bouabdallah, Daniel Burnier, Gilles Froidevaux, Xavier Greppin, Björn Jensen, Antoine Lorotte, Laetitia Mayor, Mathieu Meisser, Roland Philippsen, Ralph Piguet, Guy Ramel, Gregoire Terrien, and Nicola Tomatis. 2003. Robox at Expo.02: A large-scale installation of personal robots. *Robotics and Autonomous Systems*, 42(3-4):203–222.

Kamil Skarzynski, Marcin Stepniak, Waldemar Bartyna, and Stanislaw Ambroszkiewicz. 2017. SO-MRS: a multi-robot system architecture based on the SOA paradigm and ontology. *arXiv:1709.03300 [cs]*. ArXiv: 1709.03300.

Sebastian Thrun. 2008. Simultaneous Localization and Mapping. In Margaret E. Jefferies and Wai-Kiang Yeap, editors, *Robotics and Cognitive Approaches to Spatial Mapping*, Springer Tracts in Advanced Robotics, pages 13–41. Springer Berlin Heidelberg, Berlin, Heidelberg.

Rudolph Triebel, Kai Arras, Rachid Alami, and al. 2016. SPENCER: A Socially Aware Service Robot for Passenger Guidance and Help in Busy Airports. In David S. Wettergreen and Timothy D. Barfoot, editors, *Field and Service Robotics: Results of the 10th International Conference*, Springer Tracts in Advanced Robotics, pages 607–622. Springer International Publishing, Cham.

Barbara Tversky and Paul U. Lee. 1998. How Space Structures Language. In Christian Freksa, Christopher Habel, and Karl F. Wender, editors, *Spatial Cognition: An Interdisciplinary Approach to Representing and Processing Spatial Knowledge*, Lecture Notes in Computer Science, pages 157–175. Springer Berlin Heidelberg, Berlin, Heidelberg.

Barbara Tversky and Paul U. Lee. 1999. Pictorial and Verbal Tools for Conveying Routes. In *Spatial Information Theory. Cognitive and Computational Foundations of Geographic Information Science*, Lecture Notes in Computer Science, pages 51–64. Springer Berlin Heidelberg.

Jules Waldhart, Aurlie Clodic, and Rachid Alami. 2018. Planning Human and Robot Placements for Shared Visual Perspective. page 10.

H. Zender, O. Martnez Mozos, P. Jensfelt, G. J. M. Kruijff, and W. Burgard. 2008. Conceptual spatial representations for indoor mobile robots. *Robotics and Autonomous Systems*, 56(6):493–502.

SpatialNet: A Declarative Resource for Spatial Relations

Morgan Ulinski and **Bob Coyne** and **Julia Hirschberg**
Department of Computer Science
Columbia University
New York, NY, USA
{mulinski,coyne,julia}@cs.columbia.edu

Abstract

This paper introduces SpatialNet, a novel resource which links linguistic expressions to actual spatial configurations. SpatialNet is based on FrameNet (Ruppenhofer et al., 2016) and VigNet (Coyne et al., 2011), two resources which use frame semantics to encode lexical meaning. SpatialNet uses a deep semantic representation of spatial relations to provide a formal description of how a language expresses spatial information. This formal representation of the lexical semantics of spatial language also provides a consistent way to represent spatial meaning across multiple languages. In this paper, we describe the structure of SpatialNet, with examples from English and German. We also show how SpatialNet can be combined with other existing NLP tools to create a text-to-scene system for a language.

1 Introduction

Spatial language understanding is a research area in NLP with applications from robotics and navigation to paraphrase and image caption generation. However, most work in this area has been focused specifically on English. While there is a rich literature on the realization of spatial relations in different languages, there is no comprehensive resource which can represent spatial meaning in a formal manner for multiple languages. The development of formal models for the expression of spatial relations in different languages is a largely uninvestigated but relevant problem.

By way of motivation, consider the following translation examples. We use an NP in which we have a PP modifier, and we complete the sentence with a copula and an adjective to obtain a full sentence. The prepositions are marked in **boldface**. The English sentence is a word-for-word gloss of the German sentence except for the preposition.

In our first example, English *on* is correctly translated to German *an*:[1]

(1)   a. The painting **on** the wall is abstract.
        b. Correct translation: Das Gemälde **an** der Mauer/Wand ist abstrakt.
        c. Google Translate/Bing Translator (correct): Das Gemälde **an** der Wand ist abstrakt.

However, the correct translation changes if we are relating a cat to a wall:

(2)   a. The cat **on** the wall is grey.
        b. Correct translation: Die Katze **auf** der Mauer ist grau.
        c. Google Translate/Bing Translator (incorrect): Die Katze **an** der Wand ist grau.

The problem here is that the English preposition *on* describes two different spatial configurations: 'affixed to', in the case of the painting, and 'on top of', in the case of the cat.[2]

Similar problems appear when we translate from German to English. The painting again translates correctly:

(3)   a. Das Gemälde **an** der Mauer ist abstrakt.
        b. Correct translation: The painting **on** the wall is abstract.
        c. Google Translate/Bing Translator (correct): The painting **on** the wall is abstract.

---

[1] Note that English *wall* should be translated to *Wand* if it is a wall which has a ceiling attached to it, and *Mauer* if it is freestanding and does not help create an enclosed three-dimensional space. We ignore this particular issue in this discussion.

[2] We set aside the interpretation in which the cat is affixed to the wall similarly to a clock, which is an extraordinary interpretation and would require additional description in either language.

Proceedings of SpLU-RoboNLP, pages 61–70
Minneapolis, MN, June 6, 2019. ©2019 Association for Computational Linguistics

But when we replace the painting with the house, we no longer obtain the correct translation:

(4) a. Das Haus **an** der Mauer ist groß.
  b. Correct translation: The house **at** the wall is large/big.
  c. Google Translate (incorrect): The house **on** the wall is large.
  Bing Translator (incorrect): The house **on** the wall is big.

The problem is again that the German preposition *an* corresponds to two different spatial configurations, 'affixed to' (painting) and 'at/near' (house).

We address the issue of modeling cross-linguistic differences in the expression of spatial language by developing a deep semantic representation of spatial relations called *SpatialNet*. SpatialNet is based on two existing resources: FrameNet (Baker et al., 1998; Ruppenhofer et al., 2016), a lexical database linking semantic frames to manually annotated text, and VigNet (Coyne et al., 2011), a resource extending FrameNet by grounding abstract lexical semantics with concrete graphical relations. VigNet was developed as part of the WordsEye text-to-scene system (Coyne and Sproat, 2001). SpatialNet builds on both these resources to provide a formal description of the lexical semantics of spatial relations by linking linguistic expressions both to semantic frames and to actual spatial configurations. Because of the link to VigNet and WordsEye, SpatialNet can also be used to create a text-to-scene system for a language. This text-to-scene system can be used to verify the accuracy of a SpatialNet resource with native speakers of a language.

SpatialNet is divided into two modules: *Spatio-graphic primitives* (SGPs) represent possible graphical (spatial) relations. The *ontology* represents physical objects and their classification into semantic categories. Both are based on physical properties of the world and do not depend on a particular language. *Spatial frames* are language-specific (though, like the frames of FrameNet, may be shared among many languages) and represent the lexical meanings a language expresses. *Spatial vignettes* group together lexical items, spatial frames, and SGPs with spatial and graphical constraints from the ontology, grounding the meaning in a language-independent manner.

In Section 2, we discuss related work. In Section 3, we provide background information on FrameNet and VigNet. In Section 4, we describe the SpatialNet structure, with English and German examples. In Section 5, we show how the Spatial-Net for a language can be used in conjunction with the WordsEye text-to-scene system to generate 3D scenes from input text in that language. We conclude in Section 6 and discuss future work.

2   Related Work

Spatial relations have been studied in linguistics for many years. One study for English by Herskovits (1986) catalogs fine-grained distinctions in the interpretation of prepositions. For example, she distinguishes among the uses of *on* to mean 'on the top of a horizontal surface' (*the cup is on the table*) or 'affixed to a vertical surface' (*the picture is on the wall*). Likewise, Feist and Gentner (1998) describe user perception experiments that show that the shape, function, and animacy of the figure and ground objects are factors in the perception of spatial relations as *in* or *on*.

Other work looks at how the expression of spatial relations varies across languages. Bowerman and Choi (2003) describe how Korean linguistically differentiates between putting something in a loose-fitting container (*nehta*, e.g. fruit in a bag) vs. in a tight fitting wrapper (*kkita*, e.g. hand in glove). Other languages (English included) do not make this distinction. Levinson (2003) and colleagues have also catalogued profound differences in the ways different languages encode relations between objects in the world. Our work differs from linguistic efforts such as these in that we are building a formal representation of how a language expresses spatial information, which can be applied to a variety of NLP problems and applications. Since the representation is human- as well as machine-readable, it can also be used in more traditional linguistics.

Another area of research focuses on computational processing of spatial language. Pustejovsky (2017) has developed an annotation scheme for labeling text with spatial roles. This type of annotation can be used to train classifiers to automatically perform the task, as demonstrated by the SpaceEval task (Pustejovsky et al., 2015). Although this work provides examples of how a language expresses spatial relations, annotation of spatial roles does not provide a formal description of the link between surface realization and underlying semantics. Our work provides a formal description and also a semantic grounding that tells us the actual spatial configuration denoted by a set of spatial roles. Also, our work extends to languages other than English.

Petruck and Ellsworth (2018) advocate using FrameNet (Ruppenhofer et al., 2016) to represent spatial language. FrameNet uses frame semantics to encode lexical meaning. VigNet (Coyne et al., 2011) is an extension of FrameNet used in the WordsEye text-to-scene system (Coyne and Sproat, 2001). SpatialNet builds on both Frame-Net and VigNet; we will describe FrameNet and VigNet in more detail in the next section.

3   Background on FrameNet and VigNet

FrameNet encodes lexical meaning using a frame-semantic conceptual framework. In FrameNet, lexical items are grouped together in *frames* according to shared semantic structure. Every frame contains a number of *frame elements* (semantic roles) which are participants in this structure. Words that evoke a frame are called *lexical units*. A lexical unit is also linked to sentences that have been manually annotated to identify frame element fillers and their grammatical functions. This results in a set of *valence patterns* that represent possible mappings between syntactic functions and frame elements for the lexical unit. Frame-Net already contains a number of frames for spatial language. Spatial language frames in Frame-Net inherit from LOCATIVE-RELATION, which defines core frame elements FIGURE and GROUND, as well as non-core frame elements including DISTANCE and DIRECTION. Examples of spatial language frames are SPATIAL-CONTACT, CONTAINMENT and ADJACENCY.

VigNet, a lexical resource inspired by and based on FrameNet, was developed as part of the Words-Eye text-to-scene system. VigNet extends Frame-Net in several ways. It adds much more fine-grained frames, primarily based on differences in graphical realization. For example, the verb "wash" can be realized in many different ways, depending on whether one is washing dishes or one's hair or a car; VigNet therefore has several different wash frames. VigNet also adds graphical semantics to frames. It does this by adding primitive graphical (typically, spatial) relations between frame element fillers. These graphical relations can represent the position, orientation, size, color, texture, and poses of objects in the scene.

The graphical semantics can be thought of as a semantic grounding; it is used by WordsEye to construct and render a 3D scene. Frames augmented with graphical semantics are called *vignettes*.

The descriptions of the graphical semantics in vignettes make use of object-centric properties called *affordances* (Gibson, 1977; Norman, 1988). Affordances include any functional or physical property that allows an object to participate in actions and relations with other objects. For example, a SEAT of a chair is used to support a sitter and the INTERIOR of a box is used to hold the contents. VigNet has a rich set of spatial affordances. Some examples are CUPPED REGIONS for objects to be *in*, CANOPIES for objects to be *under*, and TOP SURFACES for objects to be *on*.

Information about the 3D objects in WordsEye is organized in VigNet into an *ontology*. The ontology is a hierarchy of semantic types with multiple inheritance. Types include both 3D objects and more general semantic concepts. For example, a particular 3D rocking chair is a sub-type of ROCKING-CHAIR.N. Every 3D object has a semantic type and is inserted into the ontology. WordsEye also includes lexicalized concepts (e.g. *chair* tied to CHAIR.N) in the ontology. The ontology includes a knowledge base of assertions that provide more information about semantic concepts. Assertions include sizes of objects and concepts, their parts, their colors, what they typically contain, what affordances they have, and information about their function. Spatial affordances and other properties can be applied to both 3D graphical objects and to more general semantic types. For example, the general semantic type CUP.N has a CUPPED REGION affordance, since this affordance is shared by all cups. A particular 3D graphical object of a cup might have a HANDLE affordance, while another might have a LID affordance, but these spatial affordances are not tied to the super-type CUP.N.

Figure 1 shows an example of two vignettes: SELF-MOTION-FROM-FRONT.R and SELF-MOTION-FROM-PORTAL.R. Both are subtypes of SELF-MOTION-FROM.R. The yellow ovals contain semantic constraints on the objects used to instantiate the frame. For example, while the relation SELF-MOTION-FROM-FRONT.R requires only that the source of the motion be a PHYSICAL-ENTITY.N, SELF-MOTION-FROM-PORTAL.R requires that the source has a

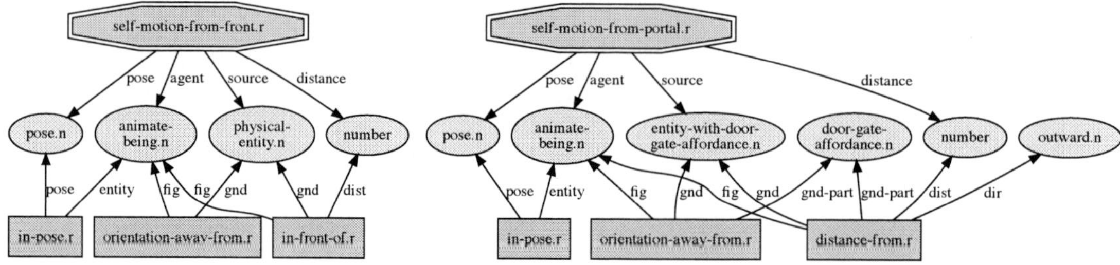

Figure 1: Two frames augmented with primitive graphical relations. The high-level semantics of SELF-MOTION-FROM-FRONT.R and SELF-MOTION-FROM-PORTAL.R are decomposed into semantic types and graphical relations.

DOOR-GATE-AFFORDANCE.N as a part.

4 Structure of SpatialNet

SpatialNet provides a formal description of spatial semantics by linking linguistic expressions to semantic frames and linking semantic frames to actual spatial configurations. To do this, we adopt some conventions from FrameNet and VigNet, making some changes to address some of the shortcomings of these resources.

FrameNet provides semantic frames including frames for spatial language. However, the syntactic information provided in the valence patterns is often insufficient for the purpose of automatically identifying frame elements in new sentences. One example is frames where the target word is a preposition, which includes many of the frames for spatial language. According to the FrameNet annotation guidelines for these (Ruppenhofer et al., 2016, page 50), the GROUND is assigned the grammatical function Obj(ect), and the FIGURE is tagged as an Ext(ernal) argument. Given a previously unseen sentence, automatic methods can identify the object of the preposition and therefore the GROUND, but the sentence may contain several noun phrases outside the prepositional phrase, making the choice of FIGURE ambiguous. FrameNet also does not provide a semantic grounding. To create SpatialNet, we adopt the concept of a FrameNet *frame*, including the definition of *frame elements* and *lexical units*. However, we modify the valence patterns to more precisely define syntactic patterns in a declarative format. In addition, to facilitating the use of SpatialNet across different languages, we specify syntactic constraints in valence patterns using labels from the Universal Dependencies project (Universal Dependencies, 2017).

VigNet does provide a grounding in graphical

semantics, but presents other problems. First, VigNet does not currently include a mapping from syntax to semantic frames. Although vignettes provide a framework for linking semantic frames to primitive graphical relations, the VigNet resource does not include frames for spatial prepositions, but only for higher-level semantic constructs. Finally, since VigNet has been developed specifically for English, some parts of the existing resource do not generalize easily to other languages. To create SpatialNet, we adopt from VigNet the concept of a *vignette* and the *semantic ontology*. However, we make the resource more applicable across languages by (a) formalizing the set of primitive graphical relations and constraints used in vignettes into what we call *spatio-graphic primitives* (SGPs), and (b) moving the language-specific mapping of lexical items to semantic categories out of the VigNet ontology and into a separate database. The SGPs and semantic ontology are used to define a language-independent semantic grounding for vignettes.

A SpatialNet for a particular language consists of a set of *spatial frames*, which link surface language to lexical semantics using valence patterns, and a set of *spatial vignettes*, which link spatial frames and lexical units to SGPs based on semantic/functional constraints. We are developing SpatialNet resources for English and German.

4.1 Ontology of Semantic Categories

The ontology in VigNet consists of a hierarchy of semantic types (concepts) and a knowledge base containing assertions. SpatialNet uses the VigNet ontology and semantic concepts directly, under the assumption that the semantic types and assertions are language-independent. Thus far, our work on English and German has not required modification of the ontology; however, since it was developed for English, it may need to be extended or modified in the future to be relevant for other languages and cultures. VigNet also includes lexicalized concepts (e.g. *chair* tied to CHAIR.N) in the ontology. For SpatialNet, we store this language-dependent lexical information in a separate database.

The mapping from lexical items to semantic concepts is important for the decomposition of text into semantics. For English SpatialNet, we use the lexical mapping extracted from VigNet. To facilitate creation of lexical mappings for other languages, we mapped VigNet concepts to entries in the Princeton WordNet of English (Princeton University, 2019). An initial mapping was constructed as follows: For each lexicalized concept in VigNet, we looked up each of its linked lexical items in WordNet. If the word (with correct part of speech) was found in WordNet, we added mappings between the VigNet concept and each WordNet synset for that word. This resulted in a many-to-many mapping of VigNet concepts to WordNet synsets. We are currently working on manually correcting this automatically-created map.

To obtain a lexical mapping for German, we use the VigNet–WordNet map in conjunction with GermaNet (Henrich and Hinrichs, 2010; Hamp and Feldweg, 1997). GermaNet includes mappings to Princeton WordNet 3.0. For a given German lexical item, we use the GermaNet links to Princeton WordNet to obtain a set of possible VigNet concepts from the VigNet–WordNet mapping. We are also experimenting with the Open German WordNet (Siegel, 2019), although in general we have found it to be less accurate. Open German WordNet includes links to the EuroWordNet Interlingual Index (ILI) (Vossen, 1998), which are in turn mapped to the Princeton English WordNet. Table 1 shows the VigNet concepts for German words used in the sentences in Figure 2, obtained using GermaNet and Open German WordNet.

4.2 Spatio-graphic Primitives

To create the set of spatio-graphic primitives used in SpatialNet, we began with relations already in VigNet. VigNet contains a range of semantic relations, from high-level abstract relations originating in FrameNet, such as ABANDONMENT.R, to low-level graphical relations, such as RGB-VALUE-OF.R. We extracted from VigNet a list of relations representing basic spatial configurations

Lexical item	VigNet concepts	
	GermaNet	**ODE-WordNet**
Mauer	WALL.N RAMPART- WALL.N RAMPART.N	WALL.N
Katze	DOMESTIC-CAT.N HOUSE-CAT.N	DOMESTIC-CAT.N HOUSE-CAT.N TRUE-CAT.N
Gemälde	PAINTING.N PICTURE.N	PICTURE.N ICON.N IMAGE.N
Haus	HOUSE.N	SHACK.N HUTCH.N HOUSE.N FAMILY.N HOME.N

Table 1: Mapping from German lexical items to VigNet semantic categories, obtained using two different German WordNet resources.

and graphical properties, separating these from the higher-level relations in VigNet which may be English-specific.

We also wanted to ensure that our list of spatio-graphic primitives was as comprehensive as possible, and not limited to the graphical capabilities of WordsEye. To that end, we annotated each picture in the Topological Relations Picture Series (Bowerman and Pederson, 1992) and the Picture Series for Positional Verbs (Ameka et al., 1999) with the spatial and graphical primitives it represents. When an appropriate spatial primitive did not exist in VigNet, we created a new one. These new SGPs have also been added it to a list of "pending" graphical relations that the WordsEye developers plan to implement in the future. In total, we have about 100 SGPs.

We use WordsEye as a realization engine for the SGPs. This is done using the WordsEye web API, which can generate a 3D scene from a semantic representation. The semantic representation consists of a list of entities, each with a semantic type from the VigNet ontology, and a list of relations between entities. SpatialNet SGPs can be used as relations in this semantic input; we are working closely with the WordsEye developers to ensure that SGPs in SpatialNet continue to be compatible with the WordsEye system. In some cases, graphical functionality for an

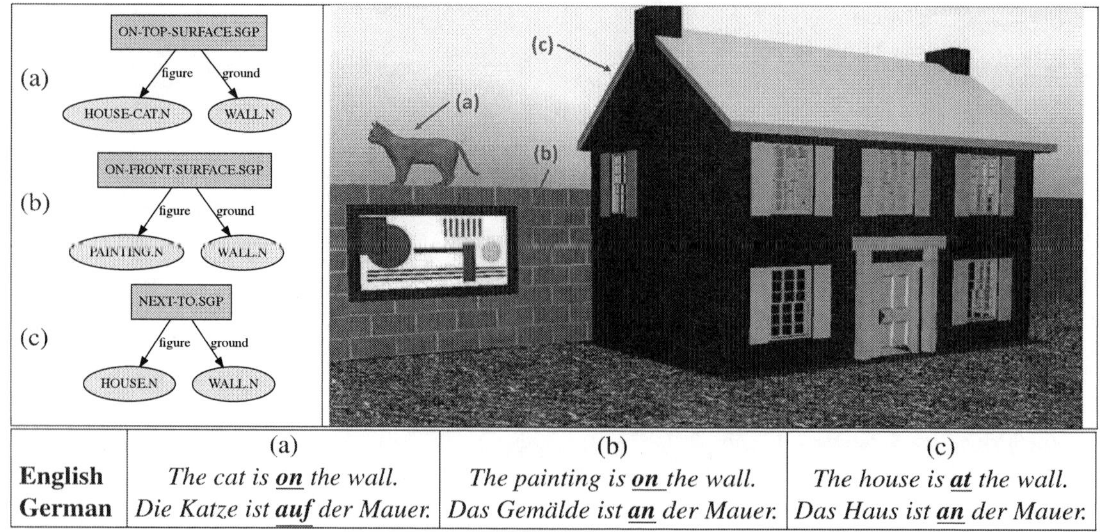

	(a)	(b)	(c)
English	*The cat is __on__ the wall.*	*The painting is __on__ the wall.*	*The house is __at__ the wall.*
German	*Die Katze ist __auf__ der Mauer.*	*Das Gemälde ist __an__ der Mauer.*	*Das Haus ist __an__ der Mauer.*

Figure 2: Examples of spatio-graphic primitives: (a) ON-TOP-SURFACE, (b) ON-FRONT-SURFACE, and (c) NEXT-TO and English/German descriptions.

SGP is not yet supported by WordsEye. For example, WordsEye currently cannot graphically represent a FITTED-ON relation, e.g. a hat on a head or a glove on a hand. When WordsEye encounters a relation that it cannot decompose into supported graphical primitives, the relation is ignored and not included in the 3D graphics. The entities referenced by the relations will be displayed in a default position (side-by-side). Figure 2 shows a scene created in WordsEye that demonstrates the spatio-graphic primitives ON-TOP-SURFACE, ON-FRONT-SURFACE, and NEXT-TO.

4.3 Spatial Frames

Spatial frames represent the lexical meanings a language can express. The structure of spatial frames is closely based on FrameNet frames. We have incorporated many of the FrameNet spatial language frames into SpatialNet, adding to these as needed. For example, for English we have added an ON-SURFACE frame that inherits from SPATIAL-CONTACT. The main difference between SpatialNet frames and FrameNet frames is in the definition of the valence patterns. SpatialNet defines valence patterns by precisely specifying lexical and syntactic constraints, which can be based on the syntactic dependency tree structure, grammatical relations, parts of speech, or lexical items. Figure 4, which provides examples of spatial vignettes, includes a valence pattern for the English lexical unit *on.adp*. This pattern specifies a syntactic structure consisting of a root (which must have part of speech NOUN), an `nsubj` dependent, and a `case` dependent (which must be the word "on"). The declarative format used to define this spatial frame is shown in Figure 3 (top).

4.4 Spatial Vignettes

Spatial vignettes use spatial frames, SGPs, and the ontology to interpret prepositions and other lexical information in a language. They relate linguistic realization (e.g. a preposition with its argument structure) to a spatial frame (such as ON-SURFACE), and at the same time to a graphical semantics expressed in terms of SGPs and additional constraints. This lexical information is often ambiguous. Consider the English and German descriptions in Figure 2. In English, the preposition *on* is ambiguous; it can mean either ON-TOP-SURFACE or ON-FRONT-SURFACE. In German, the preposition *an* is ambiguous; it can mean either ON-FRONT-SURFACE or NEXT-TO. To resolve such ambiguities, vignettes place selectional restrictions on frame elements that require fillers to have particular spatial affordances, spatial properties (such as the object size, shape, and orientation), or functional properties (such as whether the object is a vehicle or path). This information is found in the ontology.

Consider the spatial vignettes that would be used to disambiguate the meanings of English *on* from Figure 2. The declarative format used to de-

```xml
<frame name="On_surface">
  <parent name="Spatial_contact"/>
  <FE name="Figure"/>
  <FE name="Ground"/>
  <lexUnit name="on_top_of.adp">
    <pattern>
      <dep FE="Ground" tag="NOUN">
        <dep FE="Figure" reln="nsubj"/>
        <dep reln="case" word="on">
          <dep word="top" reln="mwe"/>
          <dep word="of" reln="mwe"/>
        </dep>
      </dep>
    </pattern>
  </lexUnit>

  <lexUnit name="on.adp">
    <pattern>
      <dep FE="Ground" tag="NOUN">
        <dep FE="Figure" reln="nsubj"/>
        <dep reln="case" word="on"/>
      </dep>
    </pattern>
  </lexUnit>
</frame>
```

```xml
<vignette name="on-vertical-surface">
  <input frame="On_surface"
      lexUnit="on.adp"/>
  <type-constraint FE="Ground"
      type="vertical-surface.n"/>
  <type-constraint FE="Figure"
      type="wall-item.n"/>
  <output relation="on-front-surface.r">
    <map FE="Ground" arg="ground"/>
    <map FE="Figure" arg="figure"/>
  </output>
</vignette>

<vignette name="on-top-surface">
  <input frame="On_surface"
      lexUnit="on.adp"/>
  <input frame="On_surface"
      lexUnit="on_top_of.adp"/>
  <type-constraint FE="Ground"
      type="upward-surface.n"/>
  <output relation="on-top-surface.r">
    <map FE="Ground" arg="ground"/>
    <map FE="Figure" arg="figure"/>
  </output>
</vignette>
```

Figure 3: Declarative format for spatial frames (top) and spatial vignettes (bottom)

fine these spatial vignettes is shown in Figure 3 (bottom). A visual representation of the vignettes is shown in Figure 4 (top). The vignettes link the spatial frame ON-SURFACE to different SGPs based on features of the frame element fillers.

The first vignette, ON-FRONT-SURFACE, adds semantic type constraints to both the FIGURE and the GROUND. The Figure must be of type WALL-ITEM.N and the Ground must be of type VERTICAL-SURFACE.N. If these constraints are met, the vignette produces the SGP ON-FRONT-SURFACE as output, mapping FIGURE to the SGP argument figure, and GROUND to the SGP argument ground. The second vignette, ON-TOP-SURFACE, has a semantic type constraint only that GROUND be of type UPWARD-SURFACE.N. If this constraint is met, the vignette produces the SGP ON-TOP-SURFACE. Note that, while in this case the frame elements and SGP arguments have the same names, this is not necessarily true for all vignettes (cf. the vignettes in Figure 1). Note also that in English, *painting **on** wall* is actually ambiguous, since a painting can technically be balanced on the top of a wall rather than hanging on its front surface. The spatial vignettes allow for either interpretation.

Figure 4 also shows the two vignettes which would be used to disambiguate the meanings of German *an* from Figure 2. The German vignettes link the spatial frame ADJACENCY to SGPs. The first vignette, ON-FRONT-SURFACE, is identical to the English vignette of the same name, except for the input frame and lexical unit. The semantic type constraints, SGPs, and frame element to SGP argument mappings are the same. The second vignette, NEXT-TO, does not have any semantic type constraints and thus outputs the SGP NEXT-TO with the familiar FIGURE–figure and GROUND–ground argument mappings. In the next section, we provide a complete example of using spatial vignettes to interpret these German sentences.

5 Using SpatialNet for Text-to-Scene Generation

SpatialNet can be used in conjunction with the graphics generation component of the WordsEye text-to-scene system to produce a 3D scene from a spatial description which can be used to verify the spatial frames and vignettes defined in SpatialNet. Figure 5 shows an overview of our system for text-to-scene generation. Although SpatialNet focuses on semantics, the system also requires modules for morphological analysis and syntactic parsing. For English and German, we use the Stanford CoreNLP Toolkit (Manning et al., 2014). In this section, we describe how we use Stanford CoreNLP, SpatialNet, and WordsEye to convert text into a 3D scene. We illustrate using German sentences (b) and (c) from Figure 2.

First, Stanford CoreNLP is used to perform

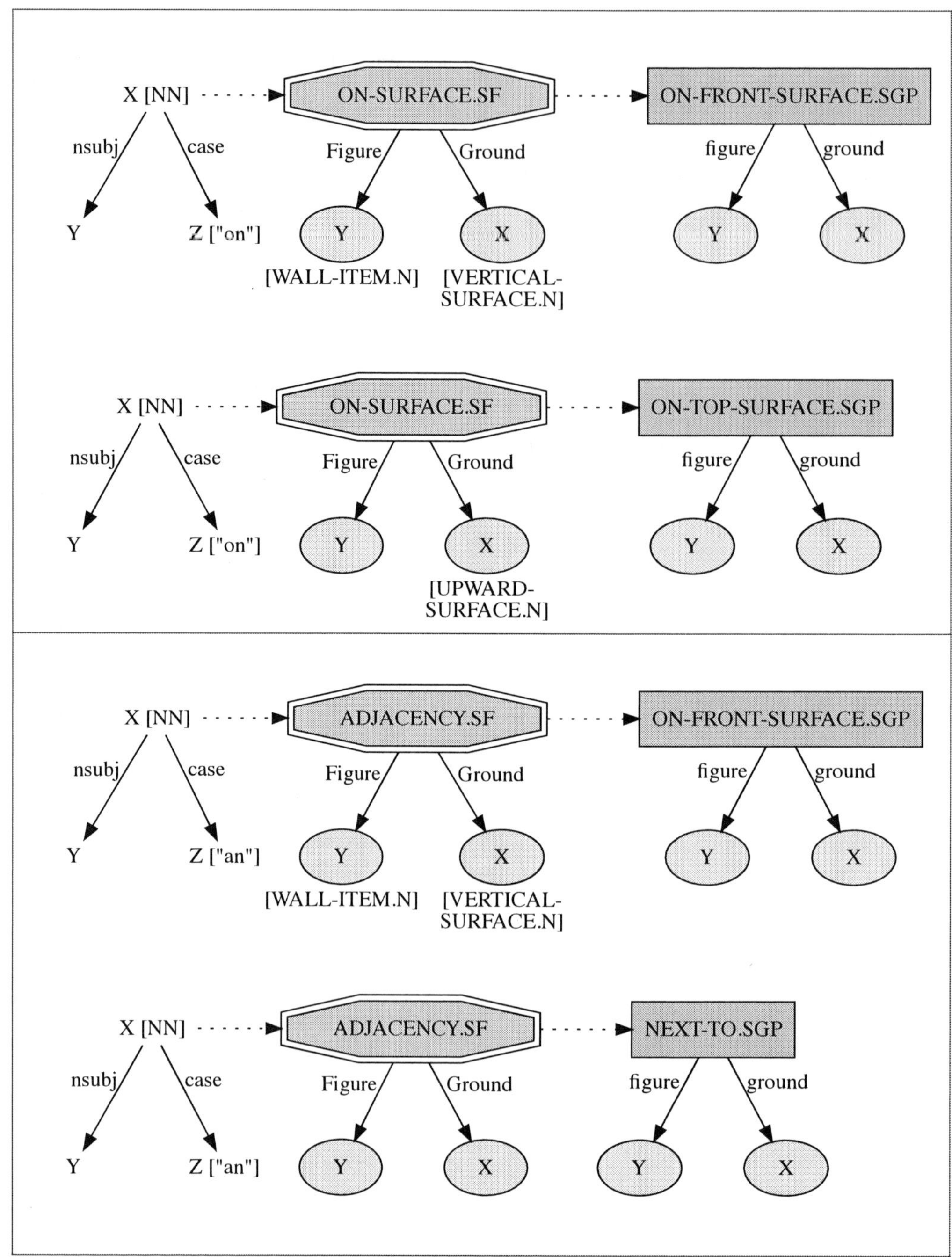

Figure 4: Spatial vignettes for different meanings of English *on* (top) and German *an* (bottom). Vignettes resolve the spatial relation given the spatial and functional object features. Spatial frames are represented by blue octagons, and SGPs by pink rectangles.

lemmatization, part-of-speech tagging, and dependency parsing. Figure 6 shows the resulting dependency structures. The dependency structures are matched against the valence patterns in spatial frames. Sentences (b) and (c) both match the valence pattern for the lexical unit *an.prep* in the ADJACENCY frame. The valence pattern identifies which lexical items in the sentence will act as frame element fillers. These lexical items are converted into semantic concepts using the lexical

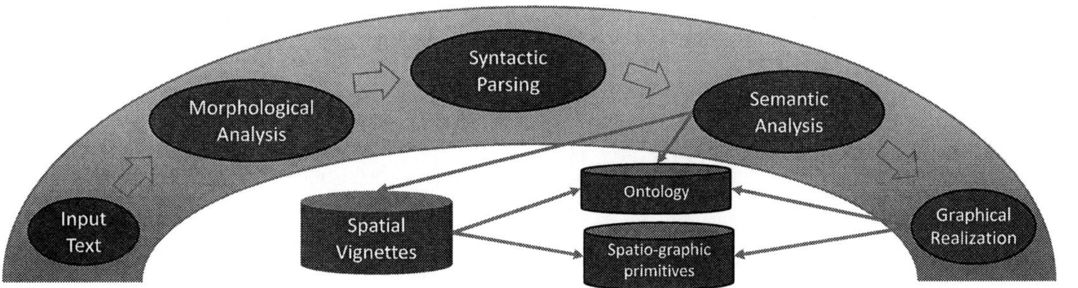

Figure 5: Pipeline for text-to-scene generation with SpatialNet

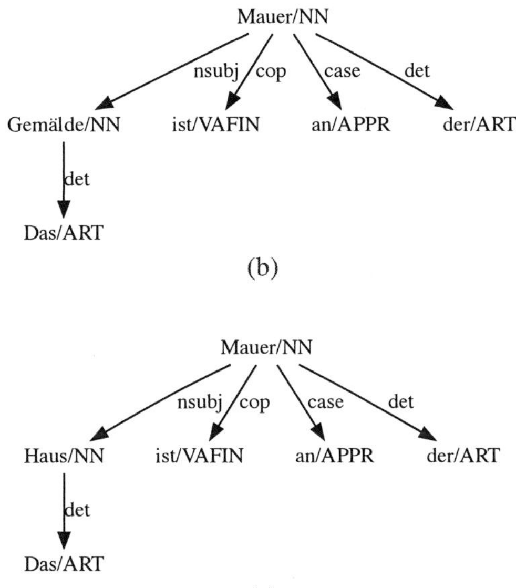

Figure 6: Results of morphological and syntactic analysis for German sentences (b) and (c)

mapping from Section 4.1. We refer to Table 1 to obtain the semantic concepts for the German lexical items. For the purposes of this example, we select the first semantic concept from the GermaNet mapping, which maps *Gemälde* to PAINTING.N, *Mauer* to WALL.N, and *Haus* to HOUSE.N.

The system then identifies the spatial vignettes which accept the frame and lexical unit as input. The features of the semantic concepts obtained for each frame element are checked against the semantic constraints in these spatial vignettes. For German sentence (b), since a WALL.N has a VERTICAL-SURFACE and a PAINTING.N is a WALL-ITEM, the ON-FRONT-SURFACE vignette is a possible match. Since a WALL.N also has an UPWARD-SURFACE, the ON-TOP-SURFACE vi-

gnette is also a possible match. For now, we select the first matching vignette, which produces the SGP ON-FRONT-SURFACE with figure PAINTING.N and ground WALL.N. For German sentence (c), since HOUSE.N is not a WALL-ITEM, only the NEXT-TO vignette is matched. This produces the SGP NEXT-TO, with figure HOUSE.N and ground WALL.N. The entities and SGPs for each sentence are then converted into a semantic representation compatible with the WordsEye web API, which is used to generate a 3D scene.

6   Summary and Future Work

We have described our development of a novel resource, SpatialNet, which provides a formal representation of how a language expresses spatial relations. We have discussed the structure of the resource, including examples from the English and German SpatialNets we are developing. We have also introduced a text-to-scene generation pipeline for using SpatialNet to convert text into 3D scenes.

In future, we will extend our semantic representation to handle motion as well as static spatial relations. A *motion vignette* could be represented by a labeled sequence of SGPs associated with key stages of the action, e.g. INITIAL-STATE, START-OF-ACTION, MIDDLE-STATE, END-OF-ACTION, FINAL-STATE. For example, *The dog jumped off the log* could be represented by the dog standing on the log, the dog leaping off with legs still on the log, the dog in mid air, the front paws touching the ground, and the dog on the ground.

In addition, we hope to extend SpatialNet to other languages, particularly low-resource and endangered languages, by incorporating it into the WordsEye Linguistics Tools (Ulinski et al., 2014a,b).

References

Felix Ameka, Carlien de Witte, and David P. Wilkins. 1999. Picture series for positional verbs: Eliciting the verbal component in locative descriptions. In David P. Wilkins, editor, *Manual for the 1999 Field Season*, pages 48–54. Max Planck Institute for Psycholinguistics, Nijmegen.

Collin F. Baker, Charles J. Fillmore, and John B. Lowe. 1998. The Berkeley FrameNet Project. In *Proceedings of the 36th Annual Meeting of the Association for Computational Linguistics and 17th International Conference on Computational Linguistics, Volume 1*, pages 86–90, Montreal, Quebec, Canada.

Melissa Bowerman and Soonja Choi. 2003. Space under Construction: Language-Specific Spatial Categorization in First Language Acquisition. In *Language in Mind: Advances in the Study of Language and Thought*, pages 387–428. MIT Press, Cambridge, MA, US.

Melissa Bowerman and Eric Pederson. 1992. Topological relations picture series. In Stephen C. Levinson, editor, *Space Stimuli Kit 1.2*, volume 51. Max Planck Institute for Psycholinguistics, Nijmegen.

Bob Coyne, Daniel Bauer, and Owen Rambow. 2011. VigNet: Grounding Language in Graphics using Frame Semantics. In *Proceedings of the ACL 2011 Workshop on Relational Models of Semantics*, pages 28–36, Portland, Oregon, USA.

Bob Coyne and Richard Sproat. 2001. WordsEye: An Automatic Text-to-scene Conversion System. In *Proceedings of the 28th Annual Conference on Computer Graphics and Interactive Techniques*, SIGGRAPH '01, pages 487–496, New York, NY, USA.

Michele I. Feist and Dedre Gentner. 1998. On Plates, Bowls, and Dishes: Factors in the Use of English IN and ON. In *Proceedings of the Twentieth Annual Meeting of the Cognitive Science Society*, pages 345–349, Hillsdale, NJ. Erlbaum.

James J. Gibson. 1977. The Theory of Affordances. In *The Ecological Approach to Visual Perception*. Erlbaum.

Birgit Hamp and Helmut Feldweg. 1997. GermaNet - a Lexical-Semantic Net for German. In *Automatic Information Extraction and Building of Lexical Semantic Resources for NLP Applications*.

Verena Henrich and Erhard Hinrichs. 2010. GernEdiT - The GermaNet Editing Tool. In *Proceedings of the Seventh Conference on International Language Resources and Evaluation (LREC'10)*, Valletta, Malta.

Annette Herskovits. 1986. *Language and Spatial Cognition: An Interdisciplinary Study of the Prepositions in English*. Cambridge University Press.

Stephen C. Levinson. 2003. *Space in Language and Cognition: Explorations in Cognitive Diversity*. Cambridge University Press, Cambridge, UK.

Christopher Manning, Mihai Surdeanu, John Bauer, Jenny Finkel, Steven Bethard, and David McClosky. 2014. The Stanford CoreNLP Natural Language Processing Toolkit. In *Proceedings of 52nd Annual Meeting of the Association for Computational Linguistics: System Demonstrations*, pages 55–60, Baltimore, Maryland.

Donald A Norman. 1988. *The Psychology of Everyday Things*. Basic Books, New York. OCLC: 874159470.

Miriam R L Petruck and Michael J Ellsworth. 2018. Representing Spatial Relations in FrameNet. In *Proceedings of the First International Workshop on Spatial Language Understanding*, pages 41–45, New Orleans. Association for Computational Linguistics.

Princeton University. 2019. WordNet: A Lexical Database for English. https://wordnet.princeton.edu/.

James Pustejovsky. 2017. ISO-Space: Annotating Static and Dynamic Spatial Information. In Nancy Ide and James Pustejovsky, editors, *Handbook of Linguistic Annotation*, pages 989–1024. Springer Netherlands, Dordrecht.

James Pustejovsky, Parisa Kordjamshidi, Marie-Francine Moens, Aaron Levine, Seth Dworman, and Zachary Yocum. 2015. SemEval-2015 Task 8: SpaceEval. In *Proceedings of the 9th International Workshop on Semantic Evaluation (SemEval 2015)*, pages 884–894, Denver, Colorado.

Josef Ruppenhofer, Michael Ellsworth, Miriam R. L Petruck, Christopher R. Johnson, Collin F. Baker, and Jan Scheffczyk. 2016. *FrameNet II: Extended Theory and Practice*.

Melanie Siegel. 2019. Open German WordNet.

Morgan Ulinski, Anusha Balakrishnan, Daniel Bauer, Bob Coyne, Julia Hirschberg, and Owen Rambow. 2014a. Documenting Endangered Languages with the WordsEye Linguistics Tool. In *Proceedings of the 2014 Workshop on the Use of Computational Methods in the Study of Endangered Languages*, pages 6–14, Baltimore, Maryland, USA.

Morgan Ulinski, Anusha Balakrishnan, Bob Coyne, Julia Hirschberg, and Owen Rambow. 2014b. WELT: Using Graphics Generation in Linguistic Fieldwork. In *Proceedings of 52nd Annual Meeting of the Association for Computational Linguistics: System Demonstrations*, pages 49–54, Baltimore, Maryland.

Universal Dependencies. 2017. Universal Dependencies. https://universaldependencies.org/.

Piek Vossen, editor. 1998. *EuroWordNet: A Multilingual Database with Lexical Semantic Networks*. Springer Netherlands.

What a neural language model tells us about spatial relations

Mehdi Ghanimifard   **Simon Dobnik**

Centre for Linguistic Theory and Studies in Probability (CLASP)
Department of Philosophy, Linguistics and Theory of Science (FLoV)
University of Gothenburg, Sweden
{mehdi.ghanimifard,simon.dobnik}@gu.se

Abstract

Understanding and generating spatial descriptions requires knowledge about *what* objects are related, their functional interactions, and *where* the objects are geometrically located. Different spatial relations have different functional and geometric bias. The wide usage of neural language models in different areas including generation of image description motivates the study of what kind of knowledge is encoded in neural language models about individual spatial relations. With the premise that the functional bias of relations is expressed in their word distributions, we construct multi-word distributional vector representations and show that these representations perform well on intrinsic semantic reasoning tasks, thus confirming our premise. A comparison of our vector representations to human semantic judgments indicates that different bias (functional or geometric) is captured in different data collection tasks which suggests that the contribution of the two meaning modalities is dynamic, related to the context of the task.

1 Introduction

Spatial descriptions such as "the chair is to the left of the table" contain spatial relations "to the left of" the semantic representations of which must be grounded in visual representations in terms of geometry (Harnad, 1990). The apprehension of spatial relations in terms of scene geometry has been investigated through acceptability scores of human judges over possible locations of objects (Logan and Sadler, 1996). In addition, other research has pointed out that there is an interplay between geometry and object-specific function in the apprehension of spatial relations (Coventry et al., 2001). Therefore, spatial descriptions must be grounded in two kinds of knowledge (Landau and Jackendoff, 1993; Coventry et al., 2001; Coventry and Garrod, 2004; Landau, 2016). One kind of knowledge is referential meaning, expressed in

the geometry of scenes (geometric knowledge or *where* objects are) while the other kind of knowledge is higher- level conceptual world knowledge about interactions between objects which is not directly grounded in perceivable situations but is learned through our experience of situations in the world (functional knowledge or *what* objects are related). Furthermore, Coventry et al. (2001) argue that individual relations have a particular geometric and functional bias and "*under*" and "*over*" are more functionally-biased than "*below*" and "*above*". For instance, when describing the relation between a person and an umbrella in a scene with a textual context such as "*an umbrella ___ a person*", "*above*" is associated with stricter geometric properties compared to "*over*" which covers a more object-specific extra-geometric sense between the target and the landmark (i.e. *covering* or *protecting* in this case). Of course, there will be several configurations of objects that could be described either with "*over*" or "*above*" which indicates that the choice of a description is determined by the speaker, in particular what aspect of meaning they want to emphasise. Coventry et al. (2001) consider this bias for prepositions that are geometrically similar and therefore the functional knowledge is reflected in different preferences for objects that are related. However, such functional differences also exist between geometrically different relations.

This poses two interesting research questions for computational modelling of spatial language. The first one is how both kinds of knowledge interact with individual spatial relations and how models of spatial language can be constructed and learned within end-to-end deep learning paradigm. Ramisa et al. (2015) compare the performance of classifiers using different multi-modal features (visual, geometric and textual) to predict a spatial preposition. Schwering (2007) applies semantic similarity metrics of spatial relations on geo-

Proceedings of SpLU-RoboNLP, pages 71–81
Minneapolis, MN, June 6, 2019. ©2019 Association for Computational Linguistics

graphical data retrieval. Collell et al. (2018) show that word embeddings can be used as predictive features for common sense knowledge about location of objects in 2D images. The second question is related to the extraction of functional knowledge for applications such as generation of spatial descriptions in a robot scenario. Typically, a robot will not be able to observe all object interactions as in (Coventry et al., 2004) to learn about the interaction of objects and choose the appropriate relation. Following the intuition that the functional bias of spatial relations is reflected in a greater selectivity for their target and landmark objects, Dobnik and Kelleher (2013, 2014) propose that the degree of association between relations and objects in the corpus of image descriptions can be used as filters for selecting the most applicable relation for a pair of objects. They also demonstrate that entropy-based analysis of the targets and landmarks can identify the functional and geometric bias of spatial relations. They use descriptions from a corpus of image descriptions because here the prepositions in spatial relations are used mainly in the spatial sense. The same investigation of textual corpora such as BNC (Consortium et al., 2007) does not yield such results as there prepositions are used mainly in their non-spatial sense.[1] Similarly, Dobnik et al. (2018) inspect the perplexity of recurrent language models for different descriptions containing spatial relations in the Visual Genome dataset of image captions (Krishna et al., 2017) in order to investigate their bias for objects.

In this paper, we follow this line of work and (i) further investigate what semantics about spatial relations are captured from descriptions of images by generative recurrent neural language models, and (ii) whether such knowledge can be extracted, for example as vector representations, and evaluated in tests. The neural embeddings are opaque to interpretations per se. The benefit of using recurrent language models is that they allow us to (i) deal with spatial relations as multi-word expressions and (ii) they learn their representations within their contexts:

(a) *a cat on a mat*
(b) *a cat on the top of a mat*
(c) *a mat under a cat*

In (a) and (b), the textual contexts are the same *"a cat ___ a mat"* but the meaning of the spatial relations, one of which is a multi-word expression, are slightly different. In (c) the context is made different through word order.

The question of what knowledge (functional or geometric) should be represented in the models can be explained in information-theoretic terms. The low surprisal of a textual language model on a new text corpora is an indication that the model has encoded the same information content as the text. In the absence of the geometric knowledge during the training of the model, this means that a language model encodes the relevant functional knowledge. We will show that the degree to which each spatial description containing a spatial relation encodes functional knowledge in different contexts can be used as source for building distributional representations. We evaluate these representations intrinsically in reasoning tests and extrinsically against human performance and human judgment.

The contributions of this paper are:

1. It is an investigation of the semantic knowledge about spatial relations learned from textual features in recurrent language models with intrinsic and extrinsic methods of evaluation on internal representations.
2. It proposes a method of inspecting contextual performance of generative neural language models over a wide categories of contexts.

This paper is organised as follows: in Section 2 we describe how we create distributional representations with recurrent neural language models, in Section 3 we describe our computational implementations that build these representations, and in Section 4 we provide their evaluation. In Section 5 we give our final remarks.

2 Neural representations of spatial relations

Distributional semantic models produce vector representations which capture latent meanings hidden in association of words in documents (Church and Hanks, 1990; Turney and Pantel, 2010). The neural word embeddings were initially introduced as a component of neural language models (Bengio et al., 2003). However, subsequently neural language models such as word2vec (Mikolov et al., 2013) and GloVe (Pennington

---

[1] We may call this metaphoric or highly functional usage which is completely absent of the geometric dimension.

et al., 2014) have become used to specifically learn word embeddings from large corpora. The word embeddings trained by these models capture world-knowledge regularities expressed in language by learning from the distribution of context words which can be used for analogical reasoning[2]. Moreover, sense embeddings (Neelakantan et al., 2014) and contextual embeddings (Peters et al., 2018) have shown to provide fine-grained representation which can discriminate between different word senses or contexts, for example in substituting synonym words and multi-words in sentences (McCarthy and Navigli, 2007).

However, meaning is also captured by generative recurrent neural language models used to generate text rather than predict word similarity. The focus of our work is to investigate what semantics about spatial relations is captured by these models. Generative language models use the chain rule of probability for step-by-step prediction of the next word in a sequence. In these models, the probability of a sequence of words (or sometimes characters) is defined as the multiplication of conditional probabilities of each word given the previous context in a sequence:

$$P(w_{1:T}) = \prod_{t=1}^{T-1} P(w_{t+1}|w_{1:t}) \qquad (1)$$

where T is the length of the word sequence. The language model estimates the probability of a sequence in Equation (1) by optimising parameters of a neural network trained over sufficient data. The internal learned parameters includes embeddings for each word token which can be used as word level representations directly.

An alternative way of extracting semantic prediction from a generative neural language model which we are going to explore in this paper is to measure the fidelity of the model's output predictions against a new ground truth sequence of words. This is expressed in the measure of *Perplexity* as follows:

$$PP(S) = (\prod_{s \in S} P(w_{1:t} = s))^{\frac{-1}{|S|}} \qquad (2)$$

where S is a collection of ground truth sentences. Perplexity is a measure of the difficulty of a gen-

eration task which is based on the information theoretic concept of entropy (Bahl et al., 1983). It is based on *cross-entropy* which takes into account the probability of a sequence of words in ground truth sentences and the probability of a language model generating that sequence. It is often used for intrinsic evaluation of word- error rates in NLP tasks (Chen et al., 1998). However, in this paper we use perplexity as a measure of fit of a pre-trained generative neural language model to a collection of sentences.

Our proposal is as follows. We start with the hypothesis that in spatial descriptions some spatial relations (those that we call functional) are more predictable from the associated word contexts of targets and landmarks than their grounding in the visual features. Hence, this will be reflected in a perplexity of a (text-based) generative language model trained on spatial descriptions. Descriptions with functionally-biased spatial relations will be easier to predict by this language model than geometrically-biased spatial descriptions and will therefore have lower perplexity. If two sequences of words where only the spatial relations differ (but target and landmark contexts as well as other words are the same) have similar perplexity, it means that such spatial relations have similar selectional requirements and are therefore similar in terms of functional and geometric bias. We can exploit this to create vector representations for spatial relations as follows. Using a dictionary of spatial relations, we extract collections of sentences containing a particular spatial relation from a held-out dataset not used in training of the language model. The collection of sentences with a particular spatial relation are our context templates. More specifically, for our list of spatial relations $\{r_1, r_2, ..., r_k\}$, we replace the original relation r_i with a target relation r_j in its collection of sentences, e.g. we replace *to the right of*$_i$ with *in front of*$_j$. The outcome is a collection of artificial sentences $S_{i \rightarrow j}$ that are identical to the human-generated sentences except that they contain a substituted spatial relation. The perplexity of the language model on these sentences represents the association between the original spatial relation and the context in which this has been projected:

$$PP(S_{i \rightarrow j}) = PP_{i,j} = P(rel_i, c_{rel_j})^{\frac{-1}{N'}} \qquad (3)$$

where c_{rel_j} is the context of rel_i, and $PP_{i,j}$ is the perplexity of the neural language model on the

---

[2]For example, "*a* is to *a** as *b* is to *b**" can be queried with simple vector arithmetic $king - man + woman \approx queen$. More specifically, with a search over vocabulary with cosine similarity: $\underset{b^* \in V/\{a^*,b,a\}}{arg\,max} \; cos(b^*, a^* - a + b)$

sentence collection where relation rel_i is artificially placed in the contexts of relation rel_j. If rel_i and rel_j are associated with similar contexts, then we expect low perplexity for $S_{i \to j}$, otherwise the perplexity will be high. Finally, the perplexity of rel_i against each collection c_{rel_j} is computed and normalised within each collection (Equation 4) and the resulting vector per rel_i over all contexts is represented as a unit vector (Equation 5).

$$m_{i,j} = \frac{PP_{i,j}}{\sum_{i'=1}^{k} PP_{i',j}} \quad (4)$$

$$\hat{v}_i = \frac{v_i}{||v_i||} \qquad v_i = (m_{i,1}, ..., m_{i,k})^T \quad (5)$$

where $\hat{v}_i$ is the vector representation of the relation rel_i. These vectors create a matrix. In a particular cell of some row and some column, high perplexity means that the spatial relation in that row is less swappable with the context in the column, while a low perplexity means that the spatial relation is highly swappable with that context. This provides a measure similar to mutual information (PPMI) in traditional distributional vectors (Church and Hanks, 1990).

In conclusion, representing multi-word spatial relations in a perplexity matrix of different contexts allows us to capture their semantics based on the predictions and the discriminatory power of the language model. If all spatial relations are equally predictable from the language model such vector representations will be identical and vector space norms will not be able to discriminate between different spatial relations. In the following sections we report on the practical details how we build the matrix (Section 3) and evaluate it on some typical semantic tasks (Section 4). The implementation and evaluation code: `https://github.com/GU-CLASP/what_nlm_srels`

3 Dataset and models

3.1 Corpus and pre-processing

We use Visual Genome region description corpus (Krishna et al., 2017). This corpus contains 5.4 million descriptions of 108 thousand images, collected from different annotators who described specific regions of each image. As stated earlier, the reason why we use a dataset of image descriptions is because we want to have spatial usages of prepositions. Other image captioning datasets such as MSCOCO (Lin et al., 2014) and Flickr30k

(Plummer et al., 2015) could also be used. However, our investigation has shown that since the task in these datasets in not to describe directly the relation between selected regions, common geometric spatial relations are almost missing in them: there are less then 30 examples for "*left of*" and "*right of*" in these datasets.

After word tokenisation with the space operator, we apply pre-processing which removes repeated descriptions per-image and also descriptions that include uncommon words with frequency less than 100[3] Then we split the sentences into 90%-10% portions. The 90% is used for training the language model (Section 3.2), and 10% is used for generating the perplexity vectors by extracting sentences with spatial relations that represent our context bins (Section 3.3). The context bins are used for generating artificial descriptions $S_{i \to j}$ on which the language model is evaluated for perplexity.

3.2 Language model and GloVe embeddings

We train a generative neural language model on the 90% of the extracted corpus (Section 3.1) which amounts to 4,537,836 descriptions of maximum length of 29 and 4,985 words in the vocabulary. We implement a recurrent language model with LSTM (Hochreiter and Schmidhuber, 1997) and a word embeddings layer similar to Gal and Ghahramani (2016) in Keras (Chollet et al., 2015) with TensorFlow (Abadi et al., 2015) as back-end. The Adam optimiser (Kingma and Ba, 2014) is used for fitting the parameters. The model is set up with 300 dimensions both for the embedding- and the LSTM units. It is trained for 20 epochs with a batch size of 1024.

In addition to the generative LSTM language model, we also train on the same corpus GloVe (VG) embeddings with 300 dimensions and a context-window of 5 words. Finally, we also use pre-trained GloVe embeddings on the Common Crawl (CC) dataset with 42B tokens[4].

---

[3] The pre-processing leaves 5,042,039 descriptions in the corpus with maximum 31 tokens per sentence. The relatively high threshold of 100 tokens is chosen to insure sufficient support in the 10% of held-out data for bucketing. We did not use OOV tokens because the goal of the evaluation is to capture object-specific properties about spatial relations and OOV tokens would interfere with this.

[4] `http://nlp.stanford.edu/data/glove.42B.300d.zip`

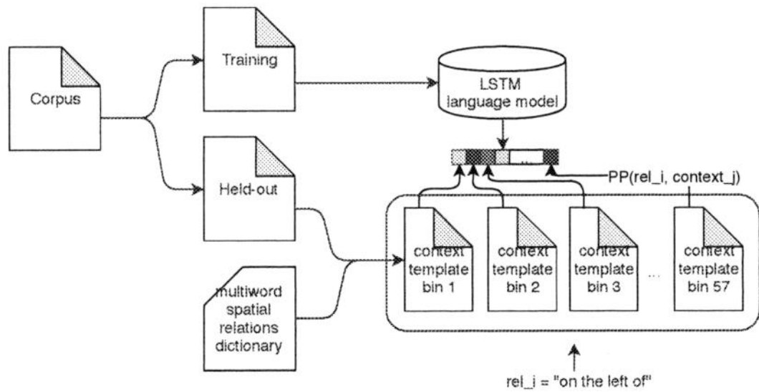

Figure 1: Generating perplexity-based vectors for each spatial relation.

3.3 Perplexity vectors

Based on the lists of spatial prepositions in (Landau, 1996) and (Herskovits, 1986), we have created a dictionary of spatial relations which include single word relations as well as all of their possible multi-word variants. This dictionary was applied on the 10% held-out dataset where we found 67 single- and multi-word spatial relation types in total. As their frequency may have fallen below 100 words due to the dataset split, we further remove all relations below this threshold which gives us 57 relations. We also create another list of relations where composite variants such as "to the left of" and "on the left of" are grouped together as "left" which contains 44 broad relations. We group the sentences by the relation they are containing to our context bins using simple pattern matching on strings. Table 1 contains some examples of our context bins. The bins are used for artificial sentence generation as explained in the previous section.

Relation (rel_i)	Context bin (c_{rel_i})
above	scissors _______ the pen
	tall building _______ the bridge
	. . .
below	pen is _______ scissors
	bench _______ the green trees
	. . .
next to	a ball-pen _______ the scissors
	car _______ the water
	. . .

Table 1: Examples of context bins based on extracted descriptions from Visual Genome. The images that belong to these descriptions are shown in Appendix B.

For each of the 67 spatial relations extracted from the larger corpus, there are 57 collections of

sentences (=the number of relations in the smaller corpus). Hence, there are $3,819 (= 67 \times 57)$ possible projections $S_{i \rightarrow j}$, where a relation i is placed in the context j, including the case where there is no swapping of relations when $j = i$. The process is shown in Figure 1. The vector of resulting perplexities in different contexts is normalised according to Equation 5 which gives us perplexity vectors (P-vectors) as shown in Figure 2.

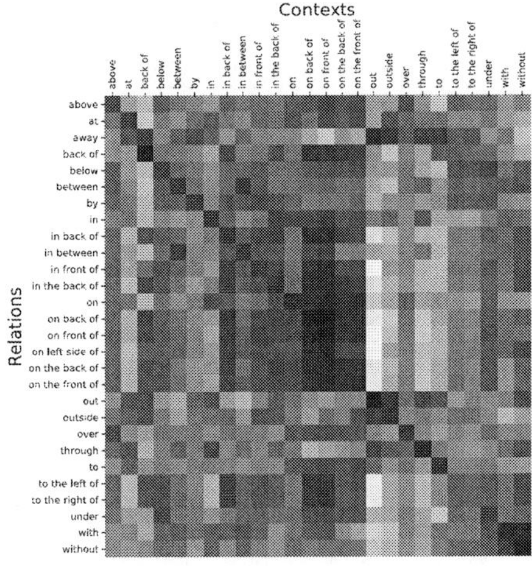

Figure 2: A matrix of perplexity vectors for 28 spatial relations and 26 contexts. For the full 67×57 matrix see Appendix C. The rows represent spatial relations and columns represent the normalised average perplexity of a language model when this relation is swapped in that context.

In addition to the P-vectors we also create representations learned by the word embedding layer in the generative language model that we train. For each of the 44 broad single-word spatial relations we extract a 300-dimensional embedding vector from the pre-trained recurrent language model (LM-vectors). In order to produce LM-vectors for the multi-word spatial relations, we simply sum the embeddings of the individual words. For example the embedding vector for "to the left of" is $v_{to} + v_{the} + v_{left} + v_{of}$. The same method is also used for the GloVe embeddings.

3.4 Human judgments

In order to evaluate our word representations we compare them to three sources of human judgments. The first one are judgments about the the fit of each spatial relation over different geometric locations of a target object in relation to a landmark which can be represented as spatial templates (Logan and Sadler, 1996). The second are 88,000 word association judgments by English speakers from (De Deyne et al., 2018). In each instance participants were presented a stimulus word and were asked to provide 3 other words. The dataset contains 4 million responses on 12,000 cues. Based on the collective performance of annotators, the dataset provides association strengths between words (which contain any kind of words, not just spatial words) as a measure of their semantic relatedness. Finally, we collected a new dataset of word similarity judgments using Amazon Mechanical Turk. Here, the participants were presented with a pair of spatial relations at a time. Their task was to use a slider bar with a numerical indicator to express how similar the pair of words are. The experiment is similar to the one described in (Logan and Sadler, 1996) except that in our case participants only saw one pair of relations at a time rather than the entire list. The shared vocabulary between these three datasets covers *left*, *right*, *above*, *over*, *below*, *under*, *near*, *next*, *away*.

4 Evaluation

As stated in Section 2 the P-vectors we have built are intended to capture the discriminatory power of a generative language model to encode and discriminate different spatial relations, their functional bias. In this section we evaluate the P-vectors on several common intrinsic and extrinsic tests for vectors. If successful, this demonstrates that such knowledge has indeed been captured by the language model. We evaluate both single- and multi-word relations.

4.1 Clustering

Method   Figure 2 and its complete version in Appendix C show that different spatial relations have different context fingerprints. To find similar relations in this matrix we can use *K-means clustering*. K-mean is a non-convex problem: different random initialisation may lead to different local minima. We apply the clustering on 67 P-vectors for multi-word spatial relations and qualitatively examine them for various sizes k. The optimal number of clusters is not so relevant here, only that for each k we get reasonable associations that follow our semantic intuitions.

Results   As shown in Table 2, with $k = 30$, the clustering of perplexity vectors shows acceptable semantics of each cluster. There are clusters with synonymous terms such as (15. *above, over*) or

1. to	18. up; down; off
2. on	19. with; without
3. away	20. together; out
4. here	21. outside; inside
5. into	22. near; beside; by
6. from	23. top; front; bottom
7. during	24. in between; between
8. back of	25. along; at; across; around
9. through	26. beneath; below; under; behind
10. alongside	27. right; back; left; side; there
11. along side	28. to the left of; to the right of; next to
12. underneath	29. in back of; in the back of; on the back of; at the top of
13. in; against	30. on the top of; on side of; on the bottom of; on left side of; on top of; on the front of; on back of; on the side of; on front of; on bottom of
14. in front of	
15. above; over	
16. to the side	
17. onto; toward	

Table 2: K-means clusters of spatial relations based on their P-vectors.

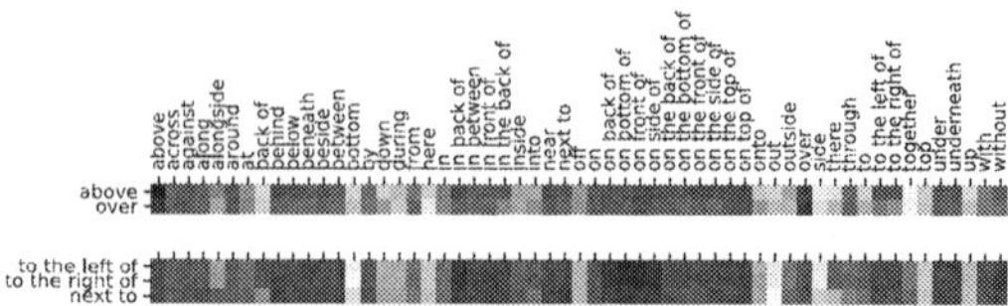

Figure 3: The P-vectors of two clusters.

(26. *below, under*). Some clusters have variants of multi-word antonymous such as (30. *on the top of, on the bottom of*). Other clusters have a mixture of such relations, e.g. (27. *right, back, left, side, and there*).

Discussion   The inspection of the perplexities of two of these clusters in Figure 3 shows that the language model has learned different selectional properties of spatial relations: *above* and *over* are generally more selective of their own contexts, while *to the left of* and *to the right of* show a higher degree of confusion with a variety of the P-vector contexts. High degree of confusion in *left* and *right* is consistent with the observation in (Dobnik and Kelleher, 2013) that these relations are less dependent on the functional relation between particular objects and therefore have a higher geometric bias. On the other hand, *above* and *over* seem to be more selective of their contexts. The functional distinction between *above* and *over* is mildly visible: the shades of blue in *above* are slightly darker than *over*.

4.2 Analogical reasoning with relations

The intrinsic properties of vector representations (the degree to which they capture functional associations between relations and their objects) can be tested with their performance in analogical reasoning tasks. We compare the performance of

	Single word	Multi-words
GloVe (CC)	0.56	0.36
GloVe (VG)	0.43	0.29
LM	0.86	0.45
P-vectors	0.62	0.47
Random	0.11	0.05

Table 3: The accuracies of different representations on the word analogy test.

the P-vectors (Section 3.3), the embeddings of the language model used to create the P-vectors and GloVe embeddings (Section 3.2) in two analogical tasks which require both geometric and functional reasoning.

4.2.1 Predicting analogical words

Method The task is similar to the analogy test (Mikolov et al., 2013; Levy et al., 2015) where two pairs of words are compared in terms of some relation "*a* is to *a'* as *b* is to *b'*". We manually grouped spatial relations that are opposite in one geometric dimension to 6 groups. These are: Group 1: left, right; Group 2: above, below; Group 3: front, back; Group 4: with, without; Group 5: in, out; and Group 6: up, down. We generate all possible permutations of these words for the analogical reasoning task which gives us 120 permutations. We expand these combinations to include multi-word variants. This dataset has 85,744 possible analogical questions such as (*above* :: *below*, *to the left of* :: ?). We accept all variants of a particular relation (e.g. *to the right side of* and *to the right of*) as the correct answer.

Results As shown in in Table 3, on the single-word test suite, the LM-embeddings perform better than other models. On multi-word test suite the P-vectors perform slightly better. On both test suites, GloVe trained on Common Crawl performs better than GloVe trained on Visual Genome. However, its performance on multi-word relations is considerably lower. We simulated random answers as a baseline to estimate the difficulty of the task. Although the multi-word test suite has ∼ 700 times more questions than the test suite with single-word relations, it is only approximately 2-times more difficult to predict the correct answer in the multi-word dataset compared to the single-word dataset.

Discussion The perplexity of the language model on complete context phrases (Multi-words) is as good indicator of semantic relatedness as the word embeddings of the underlying language

model and much better than GloVe embeddings. The good performance of the P-vectors explains the errors of the language model in generating spatial descriptions. The confusion between *in front of* and *on the back of* is similar to the confusion between *to the left of* and *to the right of* in terms of their distribution over functional contexts. Hence, a similar lack of strong functional associations allows the vectors to make inference about geometrically related word-pairs. This indicates that functional and geometric bias of words are complementary. There are two possible explanations why P-vectors perform better than LM-embeddings on multi-word vectors: (i) low-dimensions of P-vectors (57D) intensify the contribution of spatial contexts for analogical reasoning compared to high-dimensional LM-embeddings (300D); (ii) summing the vectors of the LM-embeddings for multi-words reduces their discriminatory effect.

4.2.2 Odd-one-out

Method Based on the semantic relatedness of words, the goal of this task is to find the odd member of the three. The ground truth for this test are the following five categories of spatial relations, again primarily based on geometric criteria: X-axis: left, right; Y-axis: above, over, under, below; Z-axis: front, back; Containment: in, out; and Proximity: near, away. Only the Y-axis contains words that are geometrically similar but functionally different, e.g. *above/over*. In total there are 528 possible instances with 3,456 multi-word variations. The difficulty of the task is the same for both single- and multi-word expressions as the choice is always between three words. Hence, the random baseline is 0.33.

Results Table 4 shows the accuracy in predicting the odd relation out of the three. We also add a comparison to fully geometric representations captured by spatial templates (Logan and Sadler, 1996). Ghanimifard and Dobnik (2017) show that spatial templates can be compared with Spearman's rank correlation coefficient $\rho_{X,Y}$ and therefore we also include this similarity measure. Since our groups of relations contain those that are geometric opposites in each dimension, we take the absolute value of $|\rho_{X,Y}|$. Spatial templates are not able to recognise relatedness without the right distance measure, $|\rho_{X,Y}|$. LM-embeddings perform better than other vectors in both tests, but

	Single word		Multi-words	
	$1 - cos$	$|\rho|$	$1 - cos$	$|\rho|$
GloVe (CC)	0.62	0.68	0.52	0.58
GloVe (VG)	0.61	0.61	0.58	0.59
LM	0.87	0.90	0.82	0.88
P-vectors	0.72	0.70	0.64	0.52
Sp Templates	0.22	1.0	-	-

Table 4: The accuracies in odd-one-out tests.

P-vectors follow closely. All models have a low performance on the multi-word test suite. When using $|\rho_{X,Y}|$ all vectors other than P-vectors produce better results. While we do not have an explanation for this, it is interesting to observe that $|\rho_{X,Y}|$ is a better measure of similarity than cosine.

Discussion The results demonstrate that using functional representations based on associations of words can predict considerable information about geometric distinctions between relations, e.g. distinguishing *to the right of* and *above*, and this is also true for P-vectors. As stated earlier, our explanation for this is that functional and geometric knowledge is in complementary distribution. This has positive and negative implications for joint vision and language models used in generating spatial descriptions. In the absence of geometric information, language models provide strong discriminative power in terms of functional contexts, but even if geometric latent information is expressed in them, an image captioning system still needs to ground each description in the scene geometry.

4.3 Similarity with human judgments

We compare the cosine similarity between words in LM- and P-vector spaces with similarities from (i) word association judgments (De Deyne et al., 2018), (ii) our word similarity judgments from AMT, and (iii) spatial templates (Section 3.4). We take the maximum subset of shared vocabulary between them, including *on*, *in* only shared between (i) and (ii). Since (i) is an association test, unrelated relations do not have association strengths. There are 55 total possible pairs of 11 words, while only 28 pairs are present in (i) as shown in Figure 4.

Method We take the average of the two way association strengths if the association exists and for (i) we assign a zero association for unrelated pairs such as *left* and *above*. Spearman's rank correlation coefficient $\rho_{X,Y}$ is used to compare the calculated similarities.

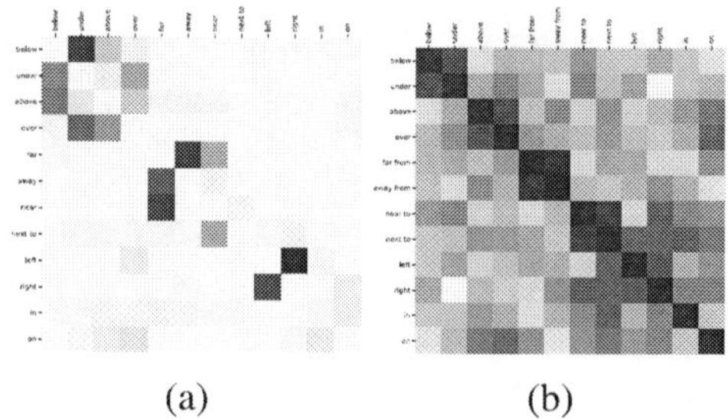

(a) (b)

Figure 4: (i) Word association judgments and (ii) word similarity judgments

Results Table 5 shows ranked correlations of different similarity measures. Spatial templates do not correlate with (WA) word associations and (WS) word similarities. On 28 pairs there is a weak negative correlation between spatial templates and WS. The correlation of similarities of two different human judgments is positive but weak ($\rho = 0.33$). The similarities predicted by LM-vectors and P-vectors correlate better with WA than WS.

	55 pairs		28 pairs	
	WA	WS	WA	WS
SpTemp	-0.02	-0.08	0.06	-0.35
LM	0.48***	0.15	0.59***	0.08
P	0.48***	0.19	0.40**	-0.08
p-values: $* < 0.01$, $** < 0.01$, $*** < 0.001$				

Table 5: Spearman's ρ between pairwise lists of similarities. WA are similarities based on word associations and WS are direct word similarities from human judgments.

Discussion The low correlation between the two similarities from human judgments is surprising. Our explanation is that this is because of different priming to functional and geometric dimension of meaning in the data collection task. In the WA task participants are not primed with the spatial domain but they are providing general word associations, hence functional associations. On the other hand, in the WS task participants are presented with two spatial relations, e.g. *left of* and *right of*, and therefore the geometric dimension of meaning is more explicitly attended. We also notice that judgments are not always unison, the same pair may be judged as similar and dissimilar which further confirms that participants are selecting between two different dimensions of meaning. This observation is consistent with our argument that LM-vectors and P-vectors encode functional knowledge. Both representations correlate better with WA than with WS. Finally, (Logan and Sadler, 1996) demonstrate that WS judgments can be decomposed to dimensions that correlate with the dimensions of the spatial templates. We leave this investigation for our future work.

5 Conclusion and future work

In the preceding discussion, we have examined what semantic knowledge about spatial relations is captured in representations of a generative neural language model. In particular, we are interested if the language model is able to encode a distinction between functional and geometric bias of spatial relations and how the two dimensions of meaning interact. The idea is based on earlier work that demonstrates that this bias can be recovered from the selectivity of spatial relations for target and landmark objects. In particular, (i) we test the difference between multi-word spatial relations at two levels: the word embeddings which are a form of internal semantic representations in a language model and the perplexity-based P-vectors which are external semantic representations based on the language model performance; (ii) we project spatial relations in the contexts of other relations and we measure the fit of the language model to these contexts using perplexity (P-vectors); (iii) we use these contexts to build a distributional model of multi-word spatial relations; (iv) in the evaluation on standard semantic similarity tasks, we demonstrate that these vectors capture fine semantic distinctions between spatial relations; (v) we also demonstrate that these representations based on word-context associations latently capture geometric knowledge that allows analogical reasoning about space; this suggests that functional and geometric components of meaning are complementary; (vi) doing so we also demonstrated that generation of spatial descriptions is also dependent on textual features, even if the system has no access to the visual features of the scene. This has implications for baselines for image captioning and how we evaluate visual grounding of spatial relations.

Our work could be extended in several ways, including by (i) using the knowledge about the bias of spatial relations to evaluate captioning tasks with spatial word substitutions (Shekhar et al., 2017a,b); (ii) examining how functional knowledge is complemented with visual knowledge in language generation (Christie et al., 2016; Delecraz et al., 2017) (iii) using different contextual embeddings such as ELMo (Peters et al., 2018) and BERT (Devlin et al., 2018) for the embedding layer of the generative language model rather than our specifically-trained word embeddings; note that P-vectors are representations of collections of context based on the performance of the decoder language model while ELMo and BERT are representations of specific context based on the encoder language model; (iv) comparing language models for spatial descriptions from different pragmatic tasks. As the focus of image captioning is to best describe the image and not for example, spatially locate a particular object, the pragmatic context of image descriptions is biased towards the functional sense of spatial relations. Our analysis should be extended to different kinds of corpora, for example those for visual question answering, human-robot interaction, and navigation instructions where we expect that precise geometric locating of objects receives more focus. Therefore, we expect to find a stronger geometric bias across all descriptions and a lower performance of our representations on analogical reasoning.

Acknowledgements

We are grateful to the anonymous reviewers for their helpful comments. The research of the authors was supported by a grant from the Swedish Research Council (VR project 2014-39) to the Centre for Linguistic Theory and Studies in Probability (CLASP) at Department of Philosophy, Linguistics and Theory of Science (FLoV), University of Gothenburg.

References

Martín Abadi, Ashish Agarwal, Paul Barham, Eugene Brevdo, Zhifeng Chen, Craig Citro, Greg S. Corrado, Andy Davis, Jeffrey Dean, Matthieu Devin, Sanjay Ghemawat, Ian Goodfellow, Andrew Harp, Geoffrey Irving, Michael Isard, Yangqing Jia, Rafal Jozefowicz, Lukasz Kaiser, Manjunath Kudlur, Josh Levenberg, Dandelion Mané, Rajat Monga, Sherry Moore, Derek Murray, Chris Olah, Mike Schuster, Jonathon Shlens, Benoit Steiner, Ilya Sutskever, Kunal Talwar, Paul Tucker, Vincent Vanhoucke, Vijay Vasudevan, Fernanda Viégas, Oriol Vinyals, Pete Warden, Martin Wattenberg, Martin Wicke, Yuan Yu, and Xiaoqiang Zheng. 2015. TensorFlow: Large-scale machine learning on heterogeneous systems. Software available from tensorflow.org.

Lalit R Bahl, Frederick Jelinek, and Robert L Mercer. 1983. A maximum likelihood approach to continuous speech recognition. *IEEE Transactions on Pattern Analysis & Machine Intelligence*, 2:179–190.

Yoshua Bengio, Réjean Ducharme, Pascal Vincent, and Christian Jauvin. 2003. A neural probabilistic language model. *Journal of machine learning research*, 3(Feb):1137–1155.

Stanley F Chen, Douglas Beeferman, and Ronald Rosenfeld. 1998. Evaluation metrics for language models. In *DARPA Broadcast News Transcription and Understanding Workshop*, pages 275–280. Citeseer.

François Chollet et al. 2015. Keras. `https://github.com/keras-team/keras`.

Gordon Christie, Ankit Laddha, Aishwarya Agrawal, Stanislaw Antol, Yash Goyal, Kevin Kochersberger, and Dhruv Batra. 2016. Resolving language and vision ambiguities together: Joint segmentation & prepositional attachment resolution in captioned scenes. *arXiv preprint arXiv:1604.02125*.

Kenneth Ward Church and Patrick Hanks. 1990. Word association norms, mutual information, and lexicography. *Computational linguistics*, 16(1):22–29.

Guillem Collell, Luc Van Gool, and Marie-Francine Moens. 2018. Acquiring common sense spatial knowledge through implicit spatial templates. In *Thirty-Second AAAI Conference on Artificial Intelligence*.

BNC Consortium et al. 2007. The british national corpus, version 3 (bnc xml edition). *Distributed by Oxford University Computing Services on behalf of the BNC Consortium*.

Kenny R Coventry, Angelo Cangelosi, Rohanna Rajapakse, Alison Bacon, Stephen Newstead, Dan Joyce, and Lynn V Richards. 2004. Spatial prepositions and vague quantifiers: Implementing the functional geometric framework. In *International Conference on Spatial Cognition*, pages 98–110. Springer.

Kenny R Coventry and Simon C Garrod. 2004. *Saying, seeing, and acting: the psychological semantics of spatial prepositions*. Psychology Press, Hove, East Sussex.

Kenny R Coventry, Mercè Prat-Sala, and Lynn Richards. 2001. The interplay between geometry and function in the comprehension of over, under, above, and below. *Journal of memory and language*, 44(3):376–398.

Simon De Deyne, Danielle J Navarro, Amy Perfors, Marc Brysbaert, and Gert Storms. 2018. The "small world of words" english word association norms for over 12,000 cue words. *Behavior research methods*, pages 1–20.

Sebastien Delecraz, Alexis Nasr, Frédéric Béchet, and Benoit Favre. 2017. Correcting prepositional phrase attachments using multimodal corpora. In *Proceedings of the 15th International Conference on Parsing Technologies*, pages 72–77.

Jacob Devlin, Ming-Wei Chang, Kenton Lee, and Kristina Toutanova. 2018. Bert: Pre-training of deep bidirectional transformers for language understanding. *arXiv preprint arXiv:1810.04805*.

Simon Dobnik, Mehdi Ghanimifard, and John D. Kelleher. 2018. Exploring the functional and geometric bias of spatial relations using neural language models. In *Proceedings of the First International Workshop on Spatial Language Understanding (SpLU 2018) at NAACL-HLT 2018*, pages 1–11, New Orleans, Louisiana, USA. Association for Computational Linguistics.

Simon Dobnik and John D. Kelleher. 2013. Towards an automatic identification of functional and geometric spatial prepositions. In *Proceedings of PRE-CogSsci 2013: Production of referring expressions – bridging the gap between cognitive and computational approaches to reference*, pages 1–6, Berlin, Germany.

Simon Dobnik and John D. Kelleher. 2014. Exploration of functional semantics of prepositions from corpora of descriptions of visual scenes. In *Proceedings of the Third V&L Net Workshop on Vision and Language*, pages 33–37, Dublin, Ireland. Dublin City University and the Association for Computational Linguistics.

Yarin Gal and Zoubin Ghahramani. 2016. A theoretically grounded application of dropout in recurrent neural networks. In *Advances in neural information processing systems*, pages 1019–1027.

Mehdi Ghanimifard and Simon Dobnik. 2017. Learning to compose spatial relations with grounded neural language models. In *Proceedings of IWCS 2017: 12th International Conference on Computational Semantics*, pages 1–12, Montpellier, France. Association for Computational Linguistics.

Stevan Harnad. 1990. The symbol grounding problem. *Physica D: Nonlinear Phenomena*, 42(1-3):335–346.

Annette Herskovits. 1986. *Language and spatial cognition: an interdisciplinary study of the prepositions in English*. Cambridge University Press, Cambridge.

Sepp Hochreiter and Jürgen Schmidhuber. 1997. Long short-term memory. *Neural computation*, 9(8):1735–1780.

Diederik P Kingma and Jimmy Ba. 2014. Adam: A method for stochastic optimization. *arXiv preprint arXiv:1412.6980*.

Ranjay Krishna, Yuke Zhu, Oliver Groth, Justin Johnson, Kenji Hata, Joshua Kravitz, Stephanie Chen, Yannis Kalantidis, Li-Jia Li, David A Shamma, et al. 2017. Visual genome: Connecting language and vision using crowdsourced dense image annotations. *International Journal of Computer Vision*, 123(1):32–73.

Barbara Landau. 1996. Multiple geometric representations of objects in languages and language learners. *Language and space*, pages 317–363.

Barbara Landau. 2016. Update on "what" and "where" in spatial language: A new division of labor for spatial terms. *Cognitive Science*, 41(2):321–350.

Barbara Landau and Ray Jackendoff. 1993. "what" and "where" in spatial language and spatial cognition. *Behavioral and Brain Sciences*, 16(2):217–238, 255–265.

Omer Levy, Yoav Goldberg, and Ido Dagan. 2015. Improving distributional similarity with lessons learned from word embeddings. *Transactions of the Association for Computational Linguistics*, 3:211–225.

Tsung-Yi Lin, Michael Maire, Serge Belongie, James Hays, Pietro Perona, Deva Ramanan, Piotr Dollár, and C Lawrence Zitnick. 2014. Microsoft coco: Common objects in context. In *European conference on computer vision*, pages 740–755. Springer.

G.D. Logan and D.D. Sadler. 1996. A computational analysis of the apprehension of spatial relations. In M. Bloom, P.and Peterson, L. Nadell, and M. Garrett, editors, *Language and Space*, pages 493–529. MIT Press.

Diana McCarthy and Roberto Navigli. 2007. Semeval-2007 task 10: English lexical substitution task. In *Proceedings of the 4th International Workshop on Semantic Evaluations*, pages 48–53. Association for Computational Linguistics.

Tomas Mikolov, Ilya Sutskever, Kai Chen, Greg S Corrado, and Jeff Dean. 2013. Distributed representations of words and phrases and their compositionality. In *Advances in neural information processing systems*, pages 3111–3119.

Arvind Neelakantan, Jeevan Shankar, Alexandre Passos, and Andrew McCallum. 2014. Efficient nonparametric estimation of multiple embeddings per word in vector space. In *Proceedings of the 2014 Conference on Empirical Methods in Natural Language Processing (EMNLP)*, pages 1059–1069.

Jeffrey Pennington, Richard Socher, and Christopher Manning. 2014. Glove: Global vectors for word representation. In *Proceedings of the 2014 conference on empirical methods in natural language processing (EMNLP)*, pages 1532–1543.

Matthew Peters, Mark Neumann, Mohit Iyyer, Matt Gardner, Christopher Clark, Kenton Lee, and Luke Zettlemoyer. 2018. Deep contextualized word representations. In *Proceedings of the 2018 Conference of the North American Chapter of the Association for Computational Linguistics: Human Language Technologies, Volume 1 (Long Papers)*, volume 1, pages 2227–2237.

Bryan A Plummer, Liwei Wang, Chris M Cervantes, Juan C Caicedo, Julia Hockenmaier, and Svetlana Lazebnik. 2015. Flickr30k entities: Collecting region-to-phrase correspondences for richer image-to-sentence models. In *Computer Vision (ICCV), 2015 IEEE International Conference on*, pages 2641–2649. IEEE.

Arnau Ramisa, Josiah Wang, Ying Lu, Emmanuel Dellandrea, Francesc Moreno-Noguer, and Robert Gaizauskas. 2015. Combining geometric, textual and visual features for predicting prepositions in image descriptions. In *Proceedings of the 2015 Conference on Empirical Methods in Natural Language Processing*, pages 214–220.

Angela Schwering. 2007. Evaluation of a semantic similarity measure for natural language spatial relations. In *International Conference on Spatial Information Theory*, pages 116–132. Springer.

Ravi Shekhar, Sandro Pezzelle, Aurélie Herbelot, Moin Nabi, Enver Sangineto, and Raffaella Bernardi. 2017a. Vision and language integration: moving beyond objects. In *IWCS 2017—12th International Conference on Computational Semantics—Short papers*.

Ravi Shekhar, Sandro Pezzelle, Yauhen Klimovich, Aurélie Herbelot, Moin Nabi, Enver Sangineto, and Raffaella Bernardi. 2017b. Foil it! find one mismatch between image and language caption. *arXiv preprint arXiv:1705.01359*.

Peter D Turney and Patrick Pantel. 2010. From frequency to meaning: Vector space models of semantics. *Journal of artificial intelligence research*, 37:141–188.

Author Index